"This fascinating, thoughtful, and challenging study tells us a great deal about what it means to be a human being."
—Karen Armstrong, *New York Times* bestselling author of
A History of God and *The Case for God*

"Lisa Miller has long been the fairest and most engaging journalist covering religion in America. In *Heaven*, she has accomplished the impossible: she has written a book about religion that fundamentalists, moderates, liberals, and nonbelievers alike can read with extreme pleasure."
—Sam Harris, *New York Times* bestselling author of *The End of Faith* and *Letter to a Christian Nation*

"[A] brainy, engaging book. . . . What Miller ultimately concludes may surprise you."   —*Entertainment Weekly*

"In this fine work, Miller . . . surveys this fascinating subject from the earliest days of Judaism to contemporary expressions of faith. Beneath her pleasing prose and often amusing observations about the afterlife, there is a longing, a desire to be part of what heaven really is. And it is this sense of personal yearning that informs her delightful and insightful study. Heaven is hope, 'a constant hope for unimaginable perfection even as we fail to achieve it.' This marvelous work is a readable and wonderfully realized study of this 'constant hope' that we share. And whether we align with Augustine or with the Mormon prophet Joseph Smith, whether we're informed by scripture or by popular culture, *Heaven* will delight and edify readers at every level."
—*Publishers Weekly* (starred review)

"With grace and insight, Lisa Miller has done a remarkable thing: she has written a wonderful book that weaves together contemporary reporting and historical scholarship with skill and energy. The result is a smart and accessible take on the ultimate question: What is Heaven? Lisa's book is a good place to begin to find an answer."

—Jon Meacham, Pulitzer Prize–winning author of *American Lion*

"A rare combination of journalism, memoir, and historical research by a self-professed skeptic who nonetheless believes in hope, this smart yet heartfelt book leads us into the center of one of the greatest conversations of all time. And Lisa Miller is the perfect conversation partner, regaling us with the wisdom (and folly) not only of the official theologies of the ancients but also of the everyday improvisations of ordinary Americans wrestling with the Big Questions forever swirling around this most audacious of hopes."

—Stephen Prothero, *New York Times* bestselling author of *American Jesus* and *Religious Literacy*

"Readers of *Heaven* will more likely than not find their own imaginations enriched, their experiences enhanced, their taste for exploration enlarged, and their impulse to reach out in empathy and hope quickened. They can then better appreciate life 'on this side' more than before. As for 'the other side,' what is ahead or beyond, honest people have to say, 'Wait and see!'"

—Martin E. Marty, Fairfax M. Cone Distinguished Service Professor Emeritus, University of Chicago

"Miller's whirlwind tour of heaven is an entertaining primer on a most complex subject." —*Booklist*

"Combines elements of journalism, academics, and memoir. Her approach provides an intriguing glimpse at what many believe the afterlife holds. . . . Miller does an exemplary job covering all monotheistic faiths. . . . Miller's book is a welcome addition. Her use of personal interviews provides a refreshing and real-life flair to her study." —*Kirkus Reviews*

HEAVEN

HEAVEN

OUR ENDURING FASCINATION
WITH THE AFTERLIFE

❧

LISA MILLER

HARPER ◖ PERENNIAL

NEW YORK • LONDON • TORONTO • SYDNEY • NEW DELHI • AUCKLAND

HARPER ● PERENNIAL

A hardcover edition of this book was published in 2010 by HarperCollins Publishers.

HarperCollins books may be purchased for educational, business, or sales promotional use. For information, please write: Special Markets Department, HarperCollins Publishers, 10 East 53rd Street, New York, NY 10022.

An extension of this copyright page appears on page 333.

FIRST HARPER PERENNIAL EDITION PUBLISHED 2011.

Designed by Eric Butler

The Library of Congress has catalogued the hardcover edition as follows:

Miller, Lisa.
    Heaven : our enduring fascination with the afterlife / Lisa Miller.—1st ed.
      p. cm.
    Includes bibliographical references (p.) and index.
    ISBN 978-0-06-055475-0
    1. Heaven. 2. Abrahamic religions. 3. Heaven—Christianity. 4. United States—Religion. I. Title.
    BL540.M56 2010
    202'.3—dc22   2009026063

ISBN 978-0-06-055476-7 (pbk.)

11 12 13 14 15 ID/RRD 10 9 8 7 6 5 4 3 2 1

For Charlie and Joey

CONTENTS

INTRODUCTION

Some years ago, I wrote a cover story for *Newsweek* magazine called "Why We Need Heaven," and the day before its publication I agreed to do a live, early-morning television interview to promote it. In the kind of ugly coincidence we journalists sometimes call lucky, events earlier that day made my story seem prescient. Ten people had been killed and fifty wounded by a Hamas "martyr" in northern Israel, and as I talked about the role that visions of heaven play in the Mideast conflict—inspiration for the suicide bombers, solace for the victims' families—images of a mangled bus and men in hazmat suits moved on the screen. I was all business, hitting my talking points. Then, as my three-minute segment was winding down, the anchor asked me a question for which I had no easy answer. I should have seen it coming.

"Do you believe in heaven?"

I was cornered. "I wish you hadn't asked me that." I smiled.

It wasn't graceful, but it was the truth. Unless we're running for public office or living on the left or right fringes of society, we Americans are as uncomfortable discussing the specifics of our spiritual lives as we are talking about the honest satisfactions and dissatisfactions of our sex lives. According to polls, most of us say we believe in heaven—and God, and miracles, and angels—but we don't think very hard about what we mean. Talk about heaven at a cocktail party—raise it as a serious topic—and watch discomfort flicker across people's faces. We say the word *heaven* out loud only when we're murmuring the Lord's Prayer; when a child asks us about the death of a pet, friend, or grandparent; or when we face death ourselves—our own or that of a loved one. Yet in spite of the pervasive sense that real talk about heaven is somehow silly, nearly all of us—me included—carry visions of heaven around in our heads. These visions have their roots in official doctrine, in creeds chanted and lessons learned at religious school. They also have roots in culture: in paintings and songs, in jokes and movies, and in the stories our parents told us. Our visions of heaven may be based as well on our own transcendent experiences—those moments in life when we see that the miracle of earthly existence is bigger and more perfect that we can conceive of day to day. How could I— journalist, religion expert, professional skeptic—tell a television news anchor that I believe I have seen the ghost of my dead grandfather? Or that I imagined heaven to be something like the heart-bursting feeling of love I had one afternoon five months after my wedding, when my husband, Charlie, and I drove to the beach after a grueling week of work and wound up at a fish shack at the end of a pier, drinking beer and squinting at boats in the sun. I've replayed that August television interview a thousand times in my head, wishing I'd handled it with more finesse. At that moment, before work on this book began in earnest, I would have said something like: "The idea of heaven defies logic, but in my mind it represents hope, and I believe in hope."

What are we talking about when we talk about heaven? We Americans inhabit perhaps the most religiously diverse society anywhere on this planet. Pluralism is written into our Constitution, it resides in our bones; even those of us who are devoutly committed to one faith regard our neighbors' rights to practice another one as sacrosanct. When he was running for president in 1960, John F. Kennedy defended his Roman Catholic faith against accusations that he was a papist. To assault someone's religion—Catholic, Quaker, or Jew—was un-American, he said: "Religious liberty is so indivisible that an act against one church is treated as an act against all." Nearly 80 percent of Americans tell pollsters they're Christian, but the uniformity stops there. Americans are liberal Protestants and conservative Protestants; they're Roman Catholic and Eastern Orthodox. (Definitions and labels, such as they are, hold only so much meaning. Write a story about Mormons calling them "Christian," and get an in-box full of e-mail from evangelicals explaining why Mormons are not Christian; write a story that distinguishes Mormons from other Christians, and your in-box is full of mail from Mormons, explaining just how Christian they are.) Americans are also non-Christian. Thanks to waves of immigration over the past 150 years from Eastern Europe, Asia, Southeast Asia, the Middle East, the West Indies, and Africa, they're also Jewish, Muslim, Hindu, Buddhist, Bahai, Zoroastrian, Santería, Jain, pagan, Wiccan, New Age, atheist, agnostic, secular, and lapsed. In America, all these religious sects live peaceably—for the most part—together.

The right to freedom of religious expression given to us by the founders protected at first a kind of parochial pluralism among Protestants, from the rational skepticism of Thomas Jefferson to the dour piety of John Adams. But the number of religions expressed in America has exploded over the past two centuries, and Americans' constitutional openness to other people's beliefs means that religious experimentation—"trying on" certain beliefs, discarding those that don't satisfy—is now widespread. According to a 2007 poll by the Pew Forum on Religion and Public Life, 28 percent of Americans practice a religion different from the one they grew up with (and if you include people who move from one Protestant denomination to another, that number is 44 percent). Also according to Pew, 65 percent of Americans believe that many different religious paths can lead to eternal salvation.

As long ago as 1831, Alexis de Tocqueville complained about what he saw as the superficiality of American religious practice. Americans, he noted, "follow a religion the way our fathers took a medicine in the month of May—if it does not do any good, people seem to say, at least it cannot do any harm." This widespread erosion of orthodoxy—bemoaned by everyone from the Dalai Lama to Pope Benedict XVI—has been hastened by the rise of virtual spiritual worlds, where people have easy access to religions other than their own, says Rodger Kamenetz, author of *The Jew in the Lotus*. Many spiritually minded Americans now feel perfectly comfortable embracing more than one religious tradition at once. Thus the Mennonite who marries a Jew and attends a progressive Episcopal church, or the Christian married to the Sikh who sends her child to an Islamic preschool. "We're no longer living in an Episcopal neighborhood or a Jewish neighborhood," Kamenetz told me once. "It's easy to look over the fence and see what the other folks are doing."

In much of Western Europe, Christianity is on the wane—the great medieval cathedrals stand empty except for tourists—and a kind of serious secularism has taken its place. In France, Belgium, the Netherlands, and the Czech Republic, about a third of people say they're atheists. America is different. We retain our religiosity—more than 90 percent of Americans say they believe in God, a percentage that has not moved significantly in sixty years—but the nature of that religiosity changes with the times. Denominations grow and diminish while other spiritual movements rise up to fill the empty places. In America, Episcopalians, Methodists, and Baptists are losing adherents, while nondenominational churches, Pentecostals, Buddhists, and people who say they belong to "no religion in particular" are gaining. Surveys such as the one by Pew are helpful but not all-encompassing, for they do not include popular but unconventional spiritual practices (yoga, 12-step programs, Scientology) and do not account for such phenomena as the five million people who purchased *The Secret*, the 2006 best seller advocating the power of positive thinking.

Such diversity presents a problem for heaven, an idea—or reality, depending on your view—that has never been characterized by conceptual consensus or clarity. Even the Bible provides no single, coherent view of the afterlife. Scholars, theologians, writers, rabbis, and poets have argued over heaven—where it is, what it looks like, who gets to go—ever since people started talking about it, at times and in places where the practice of religion was far more uniform than it is today. So when 81 percent of Americans tell pollsters that they believe in heaven, up from 72 percent ten years earlier, it's hard to know exactly what they mean—beyond an automatic and understandable hope for something after death besides the terrifying end of everything. I once dated an evangelical Christian who told me late one night in a bar over bourbon that he believed that in heaven he would see the baby Jesus. I was incredulous. Why the baby Jesus? I asked. Why not Jesus at the age of his death? Why not Jesus as a lamb, or as a king, or as a heroic warrior? My date insisted: in heaven the Christian Lord would be an infant.

My good friend Kathy, a churchgoing, progressive-minded Roman Catholic, once confided to me that her image of heaven is something like a glass-walled penthouse apartment with endless, plush white shag carpeting; when you spilled something, the carpet didn't stain. In her best seller *The Lovely Bones*, Alice Sebold paints a picture of heaven custom designed for a fourteen-year-old girl: it looks like a giant high school, surrounded by fields of flowers. Puppies and dogs of every variety frolic on the grounds and Susie, the fourteen-year-old, can eat ice cream whenever she likes. Sebold grew up in a strict Episcopalian family, going to church every Sunday, and she specifically created a heaven that had no rules: no rules for getting in, and no rules for what you did when you got there.

Before orthodox readers put this book down in disgust, be comforted. Despite our religious diversity and our rather promiscuous approach to religious identity, we Americans are also deeply conservative, and data show that our conservatism is on the rise. The most successful denominations these days are those that teach that the Bible is the infallible word of God—Assemblies of God, for example, and Jehovah's Witnesses—and require of their members a rigid adherence to certain lifestyle rules. No matter where they stand on questions of biblical infallibility, most Americans' visions of heaven are rooted in the Judeo-Christian tradition, in the Bible specifically, and in the images and interpretations of Scripture handed down through the centuries via art and culture. Almost everyone would agree that talk of heaven falls far short of what it is—or what they imagine it to be. The best accounts are only approximations. It is not my job, nor is it my intention, to prove or debunk the reality of any one vision.

When we talk about heaven, most of us mean more or less the same thing. *Heaven is a perfect place. It is the home of God, and a reward for living the right kind of life. In heaven, we live forever.* This concept of heaven was invented in Judea around 200 BCE. It changed with Christ and again with Muhammad—then with the scholars of the Middle Ages, the painters of the Renaissance, the revolutionaries of the Reformation. To America, the Puritans brought with them one kind of heaven—austere and ominous—and that heaven changed during the Civil War when 620,000 American men died, leaving their daughters, wives, sisters, and mothers without fathers and husbands, brothers and sons. It changed again with the rationalism of the early twentieth century and again with the cataclysm of World War II. In America, we are heirs to all these conceptions: when thinking about heaven, each of us unconsciously dips backward in time, we dabble, we enhance what we find there with our own histories, our own ideas of perfection. Ann Dixon, a family friend for three decades, was raised a strict Lutheran but has since abandoned any formal religious life. She lives in Vermont where she cooks and eats from her own kitchen garden. She told me that in her heaven, there would be every kind of delicious taste in the world: "cheeses," she said, "and fresh berries, and wine." We are in accord on the essence. We endlessly dispute the specifics.

Is the word *heaven* a stand-in for a supernatural mystery we will never understand, as long as we live on earth? Or is it, as Billy Graham's daughter Anne says, a real place whose boundaries we could "walk off" if we knew how to get there? Are we "ourselves" in heaven? Do we keep our bodies? Do we see our loved ones? Do we see God? Can we eat, drink, make love? How do you get there? What does it look like? A city? A garden? A static place beyond the planets, as Dante imagined, aglow with the light of God? Or is heaven some kind of abstract journey toward perfect love, perfect knowledge, or perfect truth? With fundamentalism on the rise and religion increasingly front and center in public life, some of these questions are not just existential mind-benders but are at the heart of intractable global conflict. Do Jews go to the same heaven as Muslims? Do the evangelical "saved" belong in the same place as the Easter-and-Christmas Catholics?

Excellent, even popular, books have been written about heaven before. Two standouts on the scholarly side are Colleen McDannell and Bernhard Lang's *Heaven: A History*, an overview of the Christian heaven as it changed over two thousand years; and Alan Segal's *Life*

After Death: A History of the Afterlife in the Religions of the West, an erudite and comprehensive look at the evolution of the idea of heaven from its beginnings in the Ancient Near East. For anyone interested in the history of heaven, these two volumes are indispensable. More recently, orthodox apologists have published books that defend the idea of heaven as real against what they see as the corrosive influences of atheism, secularism, and religious dilettantism. Among these: Jerry Walls, an evangelical Christian; Jeffrey Burton Russell, who is a Roman Catholic; and Jon D. Levenson, a traditionally observant Jew. On the popular side, two best sellers—Mitch Albom's *The Five People You Meet in Heaven* (more than six million copies in print) and Don Piper's *90 Minutes in Heaven* (more than three million)—show how eager mainstream readers are for ideas about heaven that might inspire or console.

This book is different. I am not a scholar, a religious apologist, or an inspirational writer; I do not aim to say definitively what heaven looks like, let alone to prove or disprove its existence. I am a journalist in the field of religion, and my goal is to write a book that might guide people through the thicket of their own views about heaven by holding up a mirror of other people's beliefs, both current and past. The book sets out on three, mostly parallel paths. First, I hope to show the dramatic range of contemporary American views of heaven through written portraits of real people—unknown and famous, orthodox and "out there"—I've met through my work and in my research for this book. Second, I hope to show how these views do and do not reflect official teaching, whatever that is. Last, I hope to show that every mind-bending question a reader might have about heaven has been asked—and answered—before, by other people in other times and places. The heaven conversation in the West is more than twenty centuries long; it's a fractious debate with common themes—these themes make up the chapter titles of this book—and no consensus forthcoming. In my mind's eye, I have started to see the people who contributed most to our modern conceptions—the authors of certain biblical passages, Augustine, Aquinas, Michelangelo, and Calvin, together with the songwriters, *New Yorker* cartoonists, and Hollywood directors—sitting around a college dining hall in black turtlenecks, parsing and pontificating on celestial matters. As always, my job is to listen to that conversation and report back on what's interesting.

I hope readers will find here ideas about heaven that are novel and surprising. A skeptic might stand back in awe, as I did, at the number of brilliant philosophers and poets through history who have passionately believed in the literal resurrection of the flesh at the end of time. A liberal Protestant might look back with appreciation at certain medieval Roman Catholics for whom "good works" mattered at least as much on the path to heaven as what today is known as "a personal relationship with Jesus Christ." Barack Obama is just this kind of person; he is Protestant but shrinks from the conventional wisdom in evangelical circles that a personal relationship with Jesus is all you need to get to heaven. "I am a big believer in not just words," then-Senator Barack Obama told my *Newsweek* colleague Richard Wolffe and me on his campaign plane, "but deeds and works." My mother-in-law, Millicent Lynn, was raised a Methodist in Minnesota. She converted to Catholicism when she married and used to think of heaven as a place in the sky where God sits on a throne. Now, after two divorces, an unhappy parting of the ways with Catholicism, and some exposure to Celtic ritual and the teachings of such New Age prophets as Marianne Williamson, she thinks of heaven as more like "a place where souls go to be with energy." She was astonished to learn that Plato and certain Orthodox Jews believed much the same thing.

Certain people throughout history—prophets, visionaries, those who have had near-death experiences—have claimed to have seen heaven and I treat these testimonials as important or even inspired stories, not factual accounts. I do not believe we know, in any empirical way, anything real about heaven. Without such evidence, the story of heaven is as much about believers as it is about belief—for how people imagine heaven changes with who they are and how they live. In the spiritual "Swing Low, Sweet Chariot," the word *home*, in the refrain "coming for to carry me home," means "heaven," of course. But it also means the town of Ripley, Ohio. Underground railroad volunteers from Ripley would help carry fugitives across the Ohio River from slavery to freedom. For American slaves, heaven was home, home was freedom, freedom was in Ripley. In 1950, at a peak of American prosperity, the evangelist Billy Graham envisaged a different heaven, a kind of bucolic, suburban dream. "We are going to sit around the fireplace and have parties and the angels will wait on us," he said, "and we'll drive down the golden streets in a yellow Cadillac convertible."

In the seventh century, Muhammad gave his parched followers, who lived in a waterless place where summer temperatures regularly spiked past a hundred degrees, an overflowing paradise brimming with rivers, fountains, and ripe fruits. More than a thousand years later, the suicide bombers on the West Bank are putting their hopes in a heaven that includes not just water and couches for lounging but seventy-two dark-eyed virgins ready to perform sexual favors for each martyr. Heaven is "a whole series of symbols," says Kevin Reinhart, associate professor of Islamic religion at Dartmouth. The seventy-two maidens are "the kind of thing you say to horny eighteen-year-olds who are going to blow themselves up."

This book focuses on the three Western monotheisms—Judaism,

Christianity, and Islam—because they share so much history and so many religious concepts—and because the topic is big enough without trying to be all-encompassing. I have given the most space to the Christian story because these images have made the biggest impact on the collective American imagination. I am not so naïve as to hope that this book will have a unifying effect. People who believe they have found the one true path to the one true heaven will likely not be persuaded to follow another, but perhaps this book will give people who are struggling to clarify what they believe about the afterlife some concepts to consider and some sense of what their traditions do and don't offer. I hope it will give even secular readers a sense of connectedness to believers in the past and provide them with an occasion for self-reflection. What people think about heaven reveals a lot about who they are.

According to the polls, more Americans believe in heaven now than they did ten years ago. A 2007 Gallup poll reported that 81 percent of Americans say they believe in heaven, up from 72 percent in 1997. That year, *Time* magazine did a cover story saying belief in heaven was dead. Preachers weren't talking about heaven, the article said; young men training to be priests didn't learn about heaven in seminary. Even evangelicals treated the topic—eternal perfection and justice—with something like embarrassment. Besides, who needed heaven when life on earth was so good? In 1997, unemployment hit a twenty-eight-year low and the Dow exceeded 7,000. *Titanic* was the movie of the year and Celine Dion swept the Grammys, winning three awards including album of the year. "Heaven," the *Time* article said, "is AWOL."

When I started covering American religion for the *Wall Street Journal* in 1998, the beat was a backwater. The religious right provided fodder for political reporters, but otherwise no one thought the religion story had any real importance; the joke was that when journalists were put out to pasture they were given either a gardening or a religion column. The next step was a gold watch and a party with free booze and boiled shrimp. American believers stayed, for the most part, safely within the parameters defined for them by the chattering classes; religion writers covered them as cultural curiosities. There were the Christian fundamentalists, who tried to influence public policy on issues like abortion and gay marriage. There were the orthodox ghettos in every denomination: Jews, Mormons, Anabaptists who receded from society to live pure lives according to Scripture. There were the "seekers" who left the faiths of their childhood to follow New Age gurus or to practice a new kind of hybrid spirituality: "JewBus"—the plural of JewBu, meaning a Jew who practiced Buddhism—were the story of the moment. Once in a while a religious person or group would make headlines for novel or incendiary behavior. The Promise Keepers brought to the country's attention the millions of mostly white, evangelical men who said they wanted to be better husbands and fathers. Jerry Falwell, always a loudmouth, sniped that "Tinky Winky" (a large, purple children's TV character) was gay. I tried to cover American religion from the perspective of regular people struggling with faith; I passionately believed then (and still believe now) that people wanted to talk about—and read about—Life's Big Questions. My editors were supportive, but in general religion was not what they would have called "a national story."

Religion—and heaven—found its permanent place on the front pages on September 11, 2001. How could it not? We Americans watched live on television as people jumped one hundred stories to their deaths, as—incredibly—those massive towers crumbled to the ground. In the days that followed we imagined the three thousand terrible deaths, one by one, as we read, with tears streaming down our faces, of last phone calls, e-mail messages, voice mails. Don't forget to pick Johnnie up from school. Jane needs new shoes. I love you. Even cynics who never gave God or heaven a second thought were driven to their knees.

We learned, with horror and disbelief, that the men responsible for this cataclysm imagined they were following instructions from God. They were martyrs, they thought, who would be rewarded after the explosion with Paradise: "You will be entering the happiest life, everlasting life," reads a handwritten note found in Mohammed Atta's luggage shortly after he piloted American Airlines 11 into the Twin Towers. "Keep that in your mind if you are plagued with a problem." My former *Wall Street Journal* colleague Paul Barrett describes sitting at his desk in the newspaper's Page One department in the weeks after the attack, trying to figure out the meanings of strange new words: *jihad, sharia, hadith.* "I'd never really thought about Islam at all," he said in a phone conversation. "I had little idea who Muslims were in this country, or how they worshipped, or how they related to other Americans." What followed were years of learning and revision by journalists like Paul and me—separating the beliefs of the radical Islamists from those American Muslims who were living next door, trying like everyone else to send their kids to college. On September 11, religion left the provinces of American cultural priorities and took its place at the center. Because the attacks caused so much needless and inexplicable death, heaven was always nearby.

There followed years of popular books, movies, and songs that alluded overtly or obliquely to death and the hope of heaven. Some were about 9/11. Many were not. There was *The Lovely Bones*, of course. In 2002, Bruce Springsteen released an album, which eventually sold two million copies, called *The Rising*, a tribute to those who died on 9/11. The title song describes Springsteen's own version of heaven, a place where children—sacred in memory—dance in a bright sky. Anne Graham Lotz, the second of the evangelist Billy Graham's five children, had already been writing a book about heaven inspired by the death of a beloved brother-in-law from a brain tumor, but after 9/11, she says, these thoughts took on new urgency. She, like everyone else in America, watched the towers fall on TV and was moved to wonder how many of the three thousand "stepped into eternity without knowing Jesus." Her book, *Heaven: My Father's House*, is an attempt to console anyone in grief with the view that a cozy, comforting heaven is a certainty, if only you know the Christian Lord.

Heaven continues to be a subject for writers and artists. In 2005, the winner of the Pulitzer Prize for fiction was *Gilead*, by Marilynne Robinson. In the novel, an aged minister writes a letter to his young son with full awareness of his imminent death, and he meditates frequently on the nature of heaven: "Boughton [the narrator's best friend, who is also old] says he has more ideas about heaven every day. He said, 'Mainly I just think about the splendors of the world and multiply by two. I'd multiply by ten or twelve if I had the energy. But two is much more than sufficient for my purposes.' So he's just sitting there, multiplying the feel of the wind by two, multiplying the smell of the grass by two."

In the academy and in the pulpit, believers, fed up with what they saw as a politically correct anything-goes approach to religion in general and heaven in particular among their colleagues and students, endeavored with works of scholarship to show that Christianity—and Judaism and Islam—were meaningless without a real belief in a real heaven. Walls, Russell, and Levenson are, in one way or another, part of this movement. More recently, the Anglican Bishop of Durham, England, N. T. Wright, wrote a book called *Surprised by Hope: Rethinking Heaven, the Resurrection and the Mission of the Church*, in which he argues that belief in Christ requires a belief that the resurrection *really happened*—the dead man came to life as God and gave salvation to all who follow him. Wright wears the purple garb of an Anglican bishop and speaks wryly of the astonishment his listeners express when he tells them they need to think about the resurrection of Jesus as a real event in history. "People have been told so often that resurrection is just a metaphor," he told my editor Jon Meacham and me in an interview for *Newsweek*. "In other words, [Jesus] went to heaven, whatever that means. And they've never realized that the word *resurrection* simply didn't mean that. If people [in the first century] had wanted to say he died and went to heaven, they had perfectly good ways of saying that."

Speaking strictly for myself, 9/11 marks a border in my psyche and memory. Everything before seems bright, silly, callow. Almost everything after—and these days more than ever—feels heavy with meaning and mortality. This change in perspective is partly due to my own changed circumstances, the inevitable effect of middle age, and the responsibilities of parenthood. I think constantly, as I never did before, of my precious blessings and of life's awful fragility. News stories about cruel or random death—little boys left to starve in Newark basements and commuter plane crashes in North Carolina—feel to me scarier, more affecting than before. In the final phases of writing this book, my friend Jerry's son, Max, died in a terrible accident. Jerry is bearing up, but his pain and grief obliterate everything else—for him, and for me when I am with him. Unthinkable things, Jerry's presence in my life reminds me, can happen to anyone at any time.

But I am not alone in my grim view. It's not just that my world feels more precarious; it's that our world *is* more precarious—with war abroad and recession at home, and everywhere getting hotter and hotter. Even before the global economic collapse of 2008, a cover line on the *Economist* magazine read "Unhappy America" and showed the Statue of Liberty in a mope. We are worried about our jobs; our savings and retirement accounts; the value of our homes; the cost of health care, child care, groceries, gas, and college. We see clearly that our children will live in a world far less hospitable than the one in which we were raised. We are worried about the fates of other people, dying from starvation and disease and AIDS. We are worried about conflicts abroad, the innocents killed. We are aging. More than 12 percent of Americans are older than sixty-five, according to the U.S. Census, a number that will increase to nearly 20 percent by 2030—and nothing focuses the mind on the afterlife, to paraphrase an old saying, like the proximity of death. According to a 2003 Pew poll, more Americans are turning to prayer than they did fifteen years ago.

The atheist writer and neuroscientist Sam Harris has it right. To believe in God—and heaven—is to believe in the supernatural. He has said it derisively: believing in God is like believing in the "flying spaghetti monster," but you can frame it another way. Belief in God requires what Samuel Taylor Coleridge called "that willing suspension of disbelief," an embrace of the unknown, the inexplicable, the mysterious. Any faithful Christian, Muslim, or Jew would agree with this statement. The story of Jesus, of the burning bush and the parting of the Red Sea, the story of the angel's visit to Muhammad asleep in his cave—the believer takes all these tales as truth with a capital T, truer than the world we live in and can see every day. So it is with heaven. Harris was moved to write *The End of Faith*, his polemic against religion and its destructive aspects, in the days after 9/11. Hard times drove him to take a hard stand against God. But other people look at the state of the world, at the many thousands of American men and women killed in Iraq and Afghanistan, at the news coverage of Chinese parents holding photos of their dead children, crushed when their flimsy schools collapsed in an earthquake, and they want to believe in supernatural help. For the sake of my friend Jerry, I want to believe in miracles. It's enough to make me pray.

Just as I was beginning to research this book, some friends invited me to dinner. "Can I ask a dumb but important question?" asked Jim, who is a well-known science writer and an atheist. Oh no, I thought. "Are you going to treat heaven as a fact? Or as a myth as it changes through history?" I took his question for what it was: a warning that to have any credibility with readers I was going to have to come clean about whether I believed in heaven. I was no more articulate with Jim than I was that morning on television.

Here are the relevant facts. I am Jewish. My maternal grandparents and my mother fled Antwerp in May 1940, on the day the Nazis began bombing the city—my mother was three years old then. I grew up in a household so assimilated that we didn't belong to a temple, didn't have bar or bat mitzvahs, didn't learn Hebrew—though we did celebrate Chanukah, Passover, and usually the High Holy Days. Unlike my father's side of the family—well-established, midwestern Jews—we did not wake up Christmas morning to a hearth crowded with presents. I married a man who was baptized a Catholic but is now a nonbeliever. We were married by an Episcopal priest. He is a good friend, and we felt that his love for us would bring more meaning to the service (which he created by melding traditional Christian and Jewish ceremonies) than a priest or rabbi with whom we had no personal connection—and it did. We consider our daughter, Josephine, to be Jewish. We belong to the Reform temple near our house and for years sent her to the school affiliated with that temple; after school, our Pentecostal babysitter tells her all about the Holy Spirit. We give presents on Chanukah, and on Christmas we put more presents under the tree.

From my mother I inherited my Jewish identity, a Judaism connected to the preparation of ceremonial foods (brisket, chicken soup) and to a profound sense of ancestry: if my grandparents hadn't escaped Europe, I wouldn't be here. From my father, a virologist at Yale University, I inherited both an impatience for insufficient explanations and a reverence for the inexplicable beauty and perfection of the natural world—and for the inspired abilities of the humans who live in it. This feeling of reverence is what I have always called God.

Like so many Americans, then, I approach religion from an uneasy, untraditional place, and like so many I have struggled with what I believe about heaven. As a child, I used to imagine I saw the face of God in the clouds in the sky, but as I grew older I found this game dissatisfying. Words like *eternity* and *bliss* hold no power for me. I happen to believe that bad people can do terrible things and get away with them, so "cosmic justice" defies my rational thinking. Any traditional vision of heaven, then—angels floating on clouds strumming harps, family reunions, visions of God, throne rooms, gated cities—fails to inspire me, though antique paintings of such visions can take my breath away. Happily, in my research for this book, I did find answers—believers whose visions made heaven seem possible or at least comprehensible, theologians and scholars whose explanations were, for me, both moving and memorable. The sources of my inspiration were unlikely because they mostly came from believers whose religious faith does not mesh neatly with my own. I think of Jon D. Levenson, the Harvard professor, who instructed me not to think too much about the mechanics of getting to heaven, but rather to put my faith in a God who can do supernatural things. I think of Father Dominic Whedbee, a Trappist monk, who lives as a recluse in central Massachusetts and prays all day long for the salvation of all the souls in the world, including mine. I think of the face on Yale professor Peter Hawkins when he described his heaven: a Bach concert that fills you up to brimming—no matter how little you know about classical music. His wineglass is by his side, the sun is setting over his balcony in downtown Boston, and his handsome face is screwed up in a silly, beaming smile. The starting point for any conversation about heaven, he reminds me, is faith.

HEAVEN

ONE

WHAT IS HEAVEN?

As we're finishing our squid salad, the priest launches into his heaven jokes. He's not wearing a clerical collar, just a casual checked shirt, the kind you'd get at Brooks Brothers, and he tells jokes with the ease of a person who has a gift for making people laugh. If you can call a priest a flirt—and I think you can—then this one is. He tells the one about Pope Benedict XVI and the dissident theologians who ascend to heaven to await their judgment. There's the one about Pope John Paul II in heaven, bargaining with God over what gift to give the faithful people on earth.

And then there's the one about the Jesuit priest and the Franciscan priest who go on a road trip to Florida and get hit by a truck. The clouds part, the pearly gates appear, and the two priests stand outside in eager anticipation of their eternal destiny. The gates open and a red carpet magically rolls out and stops at the feet of the Jesuit, who stands, gaping, as all the saints walk down the carpet and embrace him. A choir of angels begins to sing. Then a blue carpet rolls out on top of the red one, and the Blessed Virgin Mary appears. Finally, a white carpet rolls out on top of the other two, and Jesus himself walks through the gates. Together, the joyful band—the saints, the Virgin, the Lord, and the happy Jesuit—turns to enter the City of God. The gates close.

The Franciscan, bewildered, is left alone. Finally, a small wooden door in one of the jeweled walls opens and a shabby monk in a brown cassock beckons him in. "What's this?" the Franciscan asks. "I'm happy to be here and all, I'm not complaining, but my friend the Jesuit, he got red carpet and blue carpet and white carpet, he got saints and choirs of angels, and he was greeted by the Blessed Virgin and the Lord himself."

His host claps him warmly on the back and as the two walk into the holy city, he says, "That's the first Jesuit they've seen up here in fifty years."

I laugh from my gut, and my companion, who is, not surprisingly, a Jesuit, smiles a small, pleased smile. Father James Martin—Jim to his friends—knows well how crazy the idea of heaven is to anybody with a rational mind. Martin, who holds a bachelor's degree from the University of Pennsylvania's Wharton School, worked at General Electric before he decided more than two decades ago, at the age of twenty-seven, to become a priest. Martin came home from work one evening, dissatisfied and burned out, and turned on PBS to find a documentary about the famous American monk and spiritual writer Thomas Merton. He started to think seriously about giving his life to God. Now the culture editor of the Jesuit magazine *America* and the author of a spiritual memoir, *My Life with the Saints*, Martin is an urbane and media-savvy professional Catholic, one of the people journalists frequently call for perspective, or just snappy quotes, when the pope comes to town or when the Vatican issues an incomprehensible statement on a politically explosive issue—the official embrace of a Holocaust-denying bishop or a further clarification of its reaction to the U.S. sex-abuse scandal. Martin's heaven joke is funny, of course, because in the Catholic world Jesuits have the reputation of being worldlier than other priests. They tend to be irreverent and anti-authoritarian. When a priest gets in trouble for speaking out against the Vatican or teaching an unconventional view of Jesus, you can almost guarantee that he will be a Jesuit. Jesuits drink, they smoke, and they like good restaurants. The first time I heard this joke, the role of the Jesuit belonged to a lawyer.

The joke is also funny (I'm reminded of the comedian Sarah Silverman's line, "If you have to explain it, Steve, it's not funny"), because it invokes every heaven cliché we Americans know and makes them ridiculous. You have gates and clouds and choirs of angels. You have saints and the personage of God himself. You have a judgment and an inexplicable heavenly hierarchy. Above all, you have a joke-teller whose job description, you would have to assume, must include a sincere and considered belief in heaven, making fun of the whole thing. Jim Martin does believe in heaven. "It's a beautiful idea," he says. He thinks about it all the time.

When I ask him how he does imagine heaven, Martin casts around for answers. He talks, with excited anticipation, about meeting the saints, which is not surprising. Martin believes, as Catholics always have, that the saints populate heaven. Martin especially likes the early-Renaissance Fra Angelico fresco *The Just Join the Angels in Paradise*, a detail from *The Last Judgment*, in which saints clasp hands with angels and dance in a circle in a rocky, blossoming garden. He likes the saying attributed to the nineteenth-century French nun Thérèse of Lisieux, "I believe in hell, but I believe it is empty." A student of the historical Jesus, Martin hopes that in heaven a true account of Jesus's life will be revealed to him, but he says, in an abashed way, that he understands that in heaven he may no longer care to know the things he yearned to know in life. He is convinced that individual identities and love between people will be preserved in heaven. "In some way," he says, "we will be recognized and welcomed

by those we know. . . . God would not destroy or end relationships."
Jim Martin is living proof that you can believe in heaven—and that
you can believe that heaven is unbelievable at the same time.

THE HOME OF GOD

Before diving into centuries of discord over concepts of heaven, over
what kinds of bodies we will have and whether our pets can go there,
I want first to establish where we agree. What are we talking about
when we talk about heaven? God lives there, of course, and his angels
do, too. Angels have been thought to live in heaven at least since the
days of the Hebrew Bible, when God sent them to earth—notably
to Abraham and Moses, then later, in the New Testament, to Mary,
mother of Jesus—to convey His messages. These were not the winged
cherubs of Hallmark cards, but splendid-and-petrifying agents of the
Lord who prompt stammers of fear and disbelief from those who encounter them. "For beauty is nothing," the poet Rainer Maria Rilke
wrote, echoing the Old Testament authors,

> But the beginning of terror, which we are still just able to endure,
> And we are so awed because it serenely disdains
> To annihilate us. Every angel is terrifying.

Pearl gates, jeweled walls, and gold streets have entered the popular
imagination through the book of Revelation, the last (and most controversial) book of the Christian Bible, which most scholars believe was
written around 95 CE. Saint Peter standing at the gates, checking off
the names of the naughty and nice like Santa Claus, the straight man
in so many heaven jokes—this has its roots in the Gospel of Matthew,
where Jesus tells his disciple Peter that he's in charge of the church

going forward. "I will give you the keys of the kingdom of heaven." A throne room, a banquet, a wedding—these images have their seeds in the Bible, as well as in other contemporaneous writings.

Descriptors of heaven have always been the best, the highest, the most that a people could imagine. The nineteenth-century British aesthete and Anglican pastor Sydney Smith famously mused that in heaven one might "eat pâté de foie gras to the sound of trumpets." In the 1890s, when American industrialists were building great railroads, a popular folk song compared heaven to a train station. "You'll behold the Union Depot into which your train will glide; / There you'll meet the Superintendent, God the Father, God the Son," are the lyrics to "Life's Railway to Heaven."

Imam Salahuddin Muhammad converted to Islam from an historically black church when he was thirteen years old, and now he works as a chaplain both at Bard College in New York State and at Fishkill Correctional Facility nearby. When he teaches Islam to inmates, he does not describe heaven as the Qur'an does, with its fountains and running water. "We *have* water," he told me in a phone call. "We have water running all the time. I use street language for them. I say, 'Anything we can get in life, it's going to be better than that. Cadillacs, or diamonds, or money—they will be plentiful for you.' "

According to a 2002 *Newsweek* poll, 71 percent of those who say they believe in heaven conceive of it as "an actual place," and in this chapter I will explore the most important areas of agreement. It's a place you go after you die. From childhood, most of us imagine heaven as directionally "up," beyond the sky, though its exact location is a matter of much dispute. It is the home of God and the faithful. It is perfect. It looks like a garden, and maybe also a city. And though it is an actual place, it is also eternal and infinite: it exists after the world ends, even after time ends. So although heaven is a "place," earthly notions of time and space do not apply there. When we use the word *heaven*, we mean all these things—and, of course, much more.

The English language makes talking about heaven uniquely difficult. The word *heaven* in English carries all the agreed-upon meanings—a place you go after death, the home of God, perfection, eternity, and so on—plus whatever else you dream of, minus whatever you don't believe. When we say "heaven," we mush all the ancient theological meanings together. We mean the place where we live with our spirits or souls after death *and* the place we inhabit with our resurrected bodies. We mean a place that occurs at the end of the world *and* a place that exists in real time, now. This messy conflation causes agonies especially for biblical scholars and historians who wish we would take more care with our vocabulary and say "resurrection" when we mean "resurrection." When the biblical authors said "heaven," they didn't mean what we mean today. "The whole first-century conceptuality of heaven is so completely different from post-enlightenment western assumptions that one really has to dismantle the way one hears the word and begin again," the Anglican Bishop of Durham, England, N. T. Wright, wrote to me in an e-mail. In this book, I use the word *heaven* in its largest, messiest, most modern sense—and, following Wright's advice, dismantle it when I can.

The early rabbis had a different word for each heavenly concept. When they meant "the place in the sky where God lives," they used one Hebrew word—*shamayim*, which means, simply, "sky" (or, more rightly, "skies," for the word is plural). When they meant "the good place you go after you die," they used *Gan Eden*, which means, of course, the Garden of Eden. When they meant "the restored earth at the end of time," they used the phrase *olam ha-ba*, or "the world to come." *Malchut shamayim* meant "the Kingdom of God," the community of God's followers on earth. (Because observant Jews don't write or say the name of God, they used *shamayim*—skies—in this case to stand for God or heaven.)

No word in Hebrew or in English carries the most potent modern meaning of the word *heaven*: a parallel universe that exists in real time, where God lives together with his angels and the souls of our beloved departed, and which occasionally and unexpectedly intersects with and affects our world. When we tell our children that Grandma is with God in heaven, this is what we mean—a real place, where real action is going on somewhere else. When we imagine angels floating on clouds in heaven, playing harps and blowing trumpets, or our dead smiling down upon us as we go about our business, we have this meaning in mind.

Science fiction and fantasy offer more vivid and seductive embellishments. In another realm, good creatures are at war with the forces of evil, and though regular earthlings are powerless to affect the outcome, the fate of our planet and everything we treasure—from our children to our democracy—is at stake. The *Matrix* movies, the novels of Robert Heinlein, the Harry Potter books, the *Lord of the Rings* trilogy by J. R. R. Tolkien, the works of Madeleine L'Engle—all these use the parallel universe as their foundational idea. In the third *Lord of the Rings* movie, the great wizard Gandalf describes heaven as a place separated from earth by a scrim of weather. He comforts the hobbit Pippin, traumatized by battle, with this: "The gray rain-curtain of this world rolls back," he says, "and all turns to silver gloss. And then you see it . . . white shores and beyond a far green country under a swift sunrise."

"Well, that isn't so bad," says Pippin.

"No, no it isn't," Gandalf replies.

Tolkien's friend C. S. Lewis, also an Oxford scholar, wrote *The Chronicles of Narnia* as a Christian allegory. The main characters of

The Lion, the Witch and the Wardrobe, which is the first book in the series, are four English children who discover another world behind a row of fur coats in a closet. They are children because, of course, this is a children's book—but also because in Christianity children have special access to heaven. As Jesus says, "Let the little children come to me and do not stop them; for it is to such as these that the kingdom of God belongs." In that book, there's a mighty battle, in which the children fight on the side of the good. At the end, Narnia, once on the brink of destruction, turns into something like heaven. The four children are given thrones to sit on and crowns to wear. "They lived in great joy," Lewis writes, "and if they ever remembered their life in this world, it was only as one remembers a dream." (At the end of *The Last Battle*, the final book, the children are killed in a train accident and—after Narnia's destruction and restoration—actually wind up in heaven: "All their life in this world and all their adventures in Narnia had only been the cover and the title page: now at last they were beginning Chapter One of the Great Story which no one on earth has read: which goes on forever: in which every chapter is better than the one before.")

From the parallel universe, angels—creatures with God-given powers—descend to earth to interact with favored humans. Such popular movies as *Ghost* and *City of Angels* show angels (or in the case of *Ghost*, a spirit), saving special people from death or despair. In Frank Capra's *It's a Wonderful Life*, the despondent and self-hating George Bailey is rescued from suicide by Clarence, a novice angel who comes down from heaven on Christmas Eve to show poor George what life on earth would be like without him. Clarence is sent down by Saint Joseph, who has heard the prayers of George's family. In America today 31 percent of people believe that they get "definite and specific" answers to their prayers at least once a month.

At least since 600 BCE, Jews have understood daily prayer—the facing toward Jerusalem, the binding of foreheads and arms with leather straps and boxes called tefillin, the reciting of the Shema, Israel's most ancient declaration of monotheism—as an activity that connects them, literally, with the Kingdom of God. Alan Segal is the Ingeborg Rennert Professor of Jewish Studies at Barnard College in New York, a specialist in the Bible—and, especially, in biblical views of the after-life. He speaks (or reads) countless languages, ancient and modern; his undergraduate course, Life After Death, is always overenrolled. He works out of a cavelike, book-crammed office, which, he boasts, "is rel-atively large for Barnard." He holds seminars there; it's easy to imagine six or eight graduate students squeezed around the small table, drink-ing tea from the plug-in pot. Segal reminds me that even modern Jews talk about "taking on the yoke of heaven"—that is, making a direct connection to the Kingdom of God—when they pray.

Segal tells me that the Essenes, an ascetic community of Jews who lived on a desert plateau near the Dead Sea around the time of Christ, left behind records of their religious rituals and prayers. According to these documents, discovered in caves in the middle of the twentieth century and known as the Dead Sea Scrolls, the Essenes believed that when they prayed they were enacting "a replica of what's happening in heaven," says Segal. "So exact that you're never sure whether they think they're in heaven or whether the angels are down on earth." (So convinced were the Essenes that their liturgies on earth mir-rored the singing of angels in heaven that they did not excrete on the Sabbath—for that was not something the angels did. "God," explains Segal, "doesn't like the smell of human excrement." When I e-mail this last bit of information to my friend David Gates, he responds, "I never thought the Essenes were playing with a full deck anyhow.")

Even today, observant Muslims, Christians, and Jews regard liturgical prayer—as well as the singing and chanting they do during worship—as activities that connect them, in real time, with heaven. Father Eugene Romano, a Roman Catholic priest who runs a community of hermits in suburban New Jersey, said it most explicitly. I visited him one wintry afternoon. The Hermits of Bethlehem dwell in small huts surrounded by tall pines in a secluded grotto just yards from some of New Jersey's most ostentatious mansions. Father Romano told me that he thinks about heaven most often when he's saying the Lord's Prayer, something he has done tens of thousands of times since he made his first Holy Communion when he was eight years old. "Every time I celebrate mass, I believe we are celebrating the eternal banquet in heaven . . . just to say mass slowly and reverently, for the good of souls and the praise of God, is such a powerful thing."

IN THE SKY

I went to see Alan Segal because I wanted to ask him why we believe God lives in the sky. It is the most fundamental of all ideas about heaven, yet it seemed to me not at all obvious. Why not in the leaves of the trees, or in the ocean, or, as the Balinese Hindus believe, in every rock and grain of rice? Yet the image that endures—in old European paintings and contemporary greeting cards, in cartoons and in the iconic poster for the Warren Beatty movie *Heaven Can Wait*—is of heaven as a place in the firmament amid the clouds. Some medieval painters even showed Jesus ascending to heaven on a kind of invisible elevator—only his feet and ankles are there at the top of the picture; the rest of his body is presumably above the frame. When Michelangelo painted his vision of God on the ceiling of the Sistine Chapel in Rome, the Lord was, literally, directly above—so that your neck hurts to look at him—enveloped in gauze, floating against a backdrop of blue-gray sky, surrounded by angels. Spend two minutes on YouTube and you can discover dozens of homemade videos by people who have recorded the signs of God they've seen in the sky. One, a cloud in the shape of a cross, has been viewed nearly a million times.

Even the smallest children understand heaven as a place that's above and beyond the earth, both real and supernatural at once. After I wrote this sentence, I tested to see if it was true. As I was putting my daughter, Josephine, nearly five at the time, to bed, I asked her, "What is heaven?" We were lying together in the dark, and she sat up, gesticulating toward the ceiling. "It's up there," she said, "in the sky." Then she lay back down. She paused. And continued: "Heaven is farther away than outer space, but it's near outer space. It's just an inch away from outer space. God lives there." The ancient Hebrews, like Josephine, simply assumed that God was "up."

Segal thought my question strange. We had gotten take-out sandwiches at a Barnard snack bar and retreated to his office, where we sat together before a computer concordance, showing Bible verses side by side, offering them in any number of English translations as well as in Hebrew and Greek. Segal is a large man in his sixties, who started out studying to be a rabbi but is now peevish about God. (We were once having drinks at a café on the Columbia University campus, when he said, "If the God of Israel exists as the Torah describes Him, I'd just get under my desk and wait until it was over.") His face is broad and angular. When I asked him to find me evidence that the authors of the Bible thought of God as inhabiting the sky, he gazed at me sideways. "I'm not sure where you're going with this," he said.

Almost every ancient religion in the West, Segal tells me, had a primary god, and that god lived high above the earth, in the sky or, as the ancient Greeks believed, on a mountain called Olympus. More than a thousand years before Christ, the ancestors of the people we now

call Jews lived side-by-side with other people, whom the Bible calls
Canaanites. The Hebrews believed in One God, but the Canaanites,
who for centuries resembled the Hebrews in almost every other way
(their houses and their farms are, according to archaeological records,
virtually indistinguishable), believed in many, whom they worshipped
using idols. They had a deity called Ba'al, a sky god who controlled
the weather, especially rain and storms. Like the God of Abraham, he
was inexplicable and full of contradictions, both sustaining and short-
tempered, terrifying and glorious. In the Ancient Near East, where
winter rainfall was unpredictable, often accompanied by harsh storms
and followed by months of drought—and where three thousand years
ago most people were farmers—a weather god would have had the
power to give life and to take it away.

The Egyptians, the Hebrews' neighbors to the southwest, also believed that gods lived in the sky—and that the pharaohs, after death,
ascended there to join them. The Egyptian god of immortality was
Osiris, who lived among the stars in the constellation we call Orion.
"The pyramids," Segal explains, "are like giant spaceships taking
you up to the Lord." (This idea of heaven as a place in outer space
exists even today. In the movie *South Park: Bigger, Longer and Uncut*,
the animated series' goggle-eyed antihero Kenny ascends to heaven
through a star-speckled galaxy wearing his trademark parka. He arrives at something that looks like a gate, made of clouds. A loud buzzer
sounds, and Kenny falls into hell.)

In the Torah—the first five books of the Bible from Genesis to
Deuteronomy—heaven is almost always just *shamayim*, the skies. Like
my daughter, the people of the Torah understood God to live both in
and beyond the sky. Segal describes the Hebrew God as master of the
heavens and points to Genesis 14, where He is "possessor of heaven
and earth." Like Ba'al, the God of Abraham is "clearly a weather

god," Segal tells me, a creator who has the power to make storms and lift the seas. In Exodus, the Lord helps Moses and the Israelites safely cross the Red Sea: "At the blast of your nostrils the waters piled up, the floods stood up in a heap; the deeps congealed in the heart of the sea." In Exodus, Psalms, and elsewhere, God acts on behalf of his people from his abode above them, raining down manna, "the corn from heaven."

If God lives in the sky, then what people know, or think they know, about the architecture of the universe informs the way they envision heaven. In the Middle Ages, the predominant view of cosmology was founded in the ideas of Aristotle, which had been refined by the first-century astronomer Ptolemy. According to Aristotle (fourth century BCE), the earth sat at the center of a series of as many as fifty-five nested crystalline spheres. Each of the heavenly bodies—the sun and moon, and all the planets—was mounted upon one of these spheres. Encircling all the spheres was an engine-sphere called the primum mobile. So powerful was the primum mobile, it made all the others move.

For more than a thousand years, this cosmological plan, or something like it, went unquestioned, and the religious repackaged it to reinforce their theology. It was understood that God Himself inhabited an immobile sphere beyond all the others; it was He who made the planets move. A Jewish text from the first century CE describes an ascension journey by someone named Rabbi Ishmael through seven spheres, or palaces, until he arrives at last at the throne of God. In Muslim tradition, the prophet Muhammad went on a "Night Journey." He ascends through seven spheres to heaven, where he at last meets Allah, who commands Muslims to pray fifty times a day. (On his way back down, he meets Moses, who tells him that fifty times is far too many, and that he should go back and renegotiate the deal. Muhammad bargains Allah down to five.)

In Christian cosmology, there were nine spheres, and each sphere correlated both to a planet and to a species of celestial creature or a Christian virtue. So, according to diagrams drawn in the Middle Ages, the sphere of the moon was occupied by God's angels whose energy moved it around the earth. The sphere of Mercury was occupied by archangels, while cherubs lived in the sphere of the fixed stars. The outermost sphere was the Empyrean, the home of God: heaven. It was stationary, perfect, and eternal. God's love in heaven caused the planets to circulate. When Dante ascends in "Paradiso" to the highest heaven, this is where he's going: to a place of

> light that flowed as flows a river,
> pouring its golden splendor between two banks
> painted with the wondrous colors of spring.

Around the year 200 BCE, some Jews began to believe that the faithful among them would ascend to heaven where they would live as themselves with God after they died. This was a radical change. Until then, heaven, *shamayim*, was the home of God—not of people. But history and culture had begun to breed in certain Jewish sects an overwhelming sense of doom; they were having premonitions about the end of history. The Jews who wrote prophetic scripture began to talk about eternal life with God as a reward for those who were "righteous." I will explore this shift more fully in the next chapter. It's enough to say here that until 200 BCE people didn't go to heaven. After 200 BCE, some of them did.

Scholars call this view—certainty about the imminent end of the world combined with a great hope about justice for the faithful in eternity—apocalypticism. The broad outlines of apocalyptic thinking go like this. Somewhere, in another dimension, the angels of heaven are at war with the forces of evil. God's side will ultimately win a battle that will bring about the end of the world. At that time, heaven will come to earth, and everything that is bad, corrupt, and sick will be renewed and purified. The decayed corpses of the dead will come to life in a perfected state. (Just how that works is also the subject of a later chapter.) Heaven will exist on earth. This worldview took hold among some Jews in the centuries before Christ—including, probably, Jesus himself. Today, most fundamentalist Jews, Muslims, and Christians—from the Lubavitch Jews who live in tight-knit communities in Brooklyn to the post–9/11 jihadis to the fundamentalist Christians who believe President Barack Obama is the Antichrist—are motivated by their conviction that the world will end soon, and that a savior will come to a restored earth to reign in peace. At that time, as the book of Revelation puts it, "God himself will be with them; he will wipe every tear from their eyes. Death will be no more; mourning and crying and pain will be no more, for the first things have passed away."

Rivkah Slonim is a handsome woman in her forties who was raised in the Crown Heights section of Brooklyn. I rang the doorbell of her childhood home one December night when she was visiting her father, and she took me next door to her brother's house, where we spoke quietly for an hour in his elegant, if sterile, foyer. Seven of her nine children have not yet flown the nest, and the atmosphere next door, she said, was hectic. The floors in her brother's house were new and gleaming, and I sat on a small Victorian couch, wondering whether to put my water glass on the narrow windowsill to my right or on the polished wood beneath my feet. Across from me, on a wall by a staircase, was a portrait of the late Rabbi Menachem Schneerson, the charismatic leader of the Lubavitchers who died in 1994—and who some Lubavitchers continue to believe is the Messiah. (Slonim does not adhere to this minority view, she says.)

Slonim is a Lubavitcher, a member of an ultra-Orthodox group whose religious observance is at once stringent and ecstatic. She dresses modestly, in a suit jacket and a skirt below her knees. Her manner is warm and open. She runs the Chabad House at the State University of New York at Binghamton, three and a half hours away, a center for on-campus Jewish outreach known for its come-one-come-all Sabbath dinners and its Bacchanalian Purim celebrations. Slonim speaks movingly of her hope that by living a life in which she adheres as closely as possible to the 613 mitzvoth, or commandments, which as she understands it God wants every Jew to perform, her soul will achieve after death a kind of supernatural intimacy with God—a physical, romantic wholeness and a complete understanding of her life's purpose that is the hope of every Lubavitch Jew. "The ultimate reason for doing mitzvoth is to be connected to the divine," she says. "The doing of mitzvoth is a sensuous coupling between God and his people."

Among Jews, Lubavitch theology is unique and controversial. Like the Essenes, and many other apocalyptic believers since, the Lubavitchers believe the end of time is near and will be followed by the redemption of the world. The life and death of their Rebbe, who encouraged his followers to get ready for the Messiah, heightens their anticipation of the end—something Slonim says she believes in "one million percent." When the end comes, she says, paraphrasing the medieval Jewish philosopher Maimonides, "knowledge of God will fill the earth like the waters cover the seas." Until then, she and other Lubavitchers, like other religious Jews, continue to do their mitzvoth and to pray for their dead with a prayer called Kaddish. Through Kaddish, Jews believe family members on earth can help expedite a soul's eventual ascension (through purifying, Dante-like layers) to God. Depending on a person's behavior in life—the rigor of his or her adherence to the mitzvoth—each soul requires purification of a different intensity. The eleven months of Kaddish, says Slonim, is "like a dry cleaner, the time when the soul is cleansed of the stains of the world."

THE PARADISE GARDEN

All three major monotheisms—Judaism, Christianity, and Islam— hold that in the new world, when God purifies what humans have defiled, He will perfect the universe and everything in it by fusing together or reinventing heaven and earth. Jewish and Islamic traditions generally talk about this new world as a Paradise garden—in Hebrew, *Gan Eden*; in Arabic, *Jannah*—a symbolic restoration of Eden where man and woman once lived without sin together with God. According to the 2002 *Newsweek* poll, 19 percent of Americans say they imagine heaven as a garden. Christians, who inherited Jewish tradition and Scripture, use garden imagery when they talk about heaven—but they frequently use urban images, too. Christianity was brand-new when the Romans leveled its birthplace Jerusalem in 70 CE, and all the residents there suffered the psychic trauma of that destruction. Jews say that in the new world they'll get their sacred Temple back; Christians imagine "the New Jerusalem"—a new, glorious, sparkling, walled city in which Jesus himself, according to the Book of Revelation, is the temple. According to the *Newsweek* poll, 13 percent of Americans imagine heaven as a city.

The world of the Bible was, mostly, a desert world—a world of farmers who yearned for rain and feared the weather. Gardens— walled, protected from the elements and avaricious predators, abundant with ripe fruit, flowing with water, honey, oil, and wine—were the best kind of place a poor desert farmer could imagine. Indeed, a verdant and protected garden was almost beyond imagining. Garden walls and gates are crucial to the biblical imagination; the idea of heaven as the Romantic poets or the American transcendentalists had it, as untamed nature—mountain ranges and rolling fields—was possible only when humans had sufficiently built insulating walls around themselves.

In Eden, according to Genesis, God put "every tree that is pleasant to the sight and good for food," as well as a river that parted four ways, and "every animal of the field and every bird of the air." There He put the tree of knowledge, of course, which had not yet been eaten from. Abundance, perfection, innocence, a time before strife and disappointment—Eden is all of this.

Christian monks in the Middle Ages believed that the Garden of Eden existed somewhere on earth, but it was far away and impossible to attain. Through or beyond Eden was heaven. In an effort to orient Christian believers properly in the world, these monks drew maps—less for navigational purposes than to illustrate the primacy in the world of Christ and heaven. They put Eden on these maps, far off to the east (Genesis describes Eden as being in the East), sometimes behind walls, rivers, or mountain ranges. The Ebstorf map, created in the thirteenth century and destroyed in the bombing of Hannover in 1943, showed Eden beyond China, behind a ring of mountains. The Hereford map, at Hereford Cathedral, in England, drawn around 1300, is more than four feet long and five feet high and depicts a whimsical world, where monsters and dog-headed men reside on blobby continents. Palestine is disproportionately large, to accommodate all the biblical scenery. Eden is behind an impassable wall near the top; directly above, Christ sits in judgment. Eden—and heaven—exist, but you can't get there from here. Christopher Columbus thought he had found Eden or something like it when he landed in South America in 1498. "I believe that the earthly Paradise lies here, which no one can enter except by God's leave," he wrote. Three-hundred-odd years later, the reclusive American poet Emily Dickinson wrote that

> "Heaven"—is what I cannot Reach!
> The Apple on the Tree—
> Provided it do hopeless—hang—
> That—"Heaven" is—to Me!

In medieval and Byzantine art, heaven was signaled by gold paint. Flat, two-dimensional saints and angels floated in orderly rows against gold scrims. Halos designated divinity—an appropriation by Christian painters from the Greeks, via the Roman artists who used halos to crown images of their emperors. In the Renaissance, as art and philosophy turned their focus away from an idealized Christian community and toward the individual, images of heaven changed, too, and the Garden of Eden became a popular subject, with people and animals frolicking in all innocence. In Sandro Botticelli's *La Primavera* (1477–1478), what some interpreters view as the Blessed Virgin stands in the center of a lush wood on a flower-strewn meadow. Angels dance around her, free of constraining garments, their bodies seemingly weightless. The branches over their heads offer abundant fruit, within easy reach of plucking. (In this heaven, Dickinson's apple is easily attainable.) In Benozzo Gozzoli's fifteenth-century fresco in Florence's Palazzo Medici-Riccardi, Paradise looks like Tuscany, with cypress trees in rows, birds flitting among rocky outcroppings, flowering trees and berry bushes in bloom. One of God's angels has wings like a peacock's tail. "Paradise," said the Florentine statesman Lorenzo de' Medici, "means nothing other than a most pleasant garden, abundant with all pleasing and delightful things."

This image, a lush and peaceful garden, protected from the corrosive influences of the world, where people live in harmony and innocence, still lingers in popular concepts of heaven. In *Just Like Heaven*, a 2005 movie starring Reese Witherspoon and Mark Ruffalo, a heartbroken landscape architect falls in love with the lively spirit of a woman whose comatose body is near death in the hospital. Near the end of the film, he restores her to consciousness with a kiss, but she cannot remember who he is. He jogs her memory—and resuscitates her love—by building her a rooftop garden, canopied, blossoming, and abundant. The garden is heaven on earth, a place away from the world where this contemporary Adam and Eve can pick up where they left off, wiser from their encounters with mortality, but existing in all innocence again.

The word *paradise* has its roots in the word *pairidaeza*, which means "walled garden" in the ancient language of the Persian priesthood, and many ancient cultures—the Greeks and the Egyptians—talked about a safe and fertile place where certain people would go after death. In the fifth century BCE, the Greek poet Pindar talked about this place as the Isle of the Blessed:

> There flowers of gold shine like flame,
> Some on bright trees on the land,
> Some nourished by the sea.

"Paradise" continues to hold connotations of safety and abundance. Paradise Valley is one of Arizona's most affluent communities, with an average household income of over $150,000 a year; Paradise Garden is one of the country's largest mail-order flower-bulb business, and also the name of a restaurant-nightclub in Sheepshead Bay, Brooklyn, where people from Russia get married and hold boisterous parties with much drunkenness and sparkly outfits. It shares its name with the all-you-can-eat buffet at the Flamingo Hotel in Las Vegas: there diners feast on a fresh fruit bar, a salad bar, prime rib, and littleneck clams as they look out of floor-to-ceiling glass panes at flamingos playing among waterfalls.

Nowhere is the idea of heaven as a paradise garden more important than in Islam, established in the seventh century CE in one of the hottest, driest, most inhospitable spots in the world. No wonder, then, that the Qur'an, Islam's holy book, promises that after death, the faithful will go to a garden. They will inhabit "gardens beneath which rivers flow," and rest among fountains "gushing in torrents at their command." There are four rivers in the Islamic paradise, one of milk, one of honey, one of wine, and one of water. As befits a religion founded in a place of relentless heat, Islam promises that in paradise, food won't spoil. The paradise gardens produce a variety of fruits, but especially pomegranates. (In some mystical Jewish traditions, heaven is also described as a garden of pomegranates, and when the Swiss psychologist Carl Jung had a near-death experience, he imagined himself to be in a pomegranate garden, too. Pomegranates—indigenous to the Middle East—are significant because their dramatic red color symbolizes blood, and their many seeds fertility.)

In the Muslim paradise, according to the Qu'ran, there is wine (forbidden to Muslims on earth), but it doesn't make you drunk. Men who have spent their lives toiling beneath a burning sun or brutally fighting over small patches of sand will greet each other with the word salaam, or "peace." They will wear gold bracelets and green robes of silk and recline upon upholstered couches while waiters pass goblets of cool, thirst-quenching beverages. Sensual pleasures of every sort will be granted in Paradise, not least among them the attentions of the houris—dark-eyed, full-breasted spirit women, who live "confined to pavilions . . . undefiled before them by humans or Jinn." ("Jinn," according to the Qur'an, are spirits made by Allah from fire, neither angels nor demons.) The Hebrew Bible and the New Testament may be vague about heaven, but the Qur'an is not.

THE HEAVENLY CITY

Fighting for parity in our imaginations is the idea of "the New Jerusalem," the celestial city. In 1987, the actress Diane Keaton made a strange little movie called *Heaven*, in which she interviewed dozens of people about what they imagined heaven to look like. A surprising number—surprising to me, because "city" never enters my own notions of heaven—cited urban places. "Like L.A., New York, Chicago," said one. "Seven million times bigger than New York City," said another. The conservative Roman Catholic priest Father Richard John Neuhaus, who passed away in 2009, liked to imagine heaven as something like Manhattan. An intellectual and an aesthete devoted to his many friends, Neuhaus adored his hometown. "I have sometimes suggested," he once wrote, "that over the heavenly gates will be a sign: 'From the Wonderful People Who Brought You New York City, the New Jerusalem.' . . . I add that those who in this life did not like New York City will have another place to go."

Jerusalem was the center of Jewish (and Christian) life from at least 600 BCE until 70 CE, when the Romans destroyed it and dispersed the people who lived there. It was not a large town, or politically very important. But at its center, on a high hill, the same hill where Abraham reportedly drew his knife to slay his beloved son Isaac, was the Temple. It was the most sacred place in Judaism, the province of the highest priests, site of the only altar on which Jews were permitted by law to make their sacrifices. Its destruction—not once, but twice, first in 586 BCE and then in 70 CE—created psychic wounds from which Jews still suffer today. The heavenly city promised in the book of Revelation is Jerusalem, more magnificent than it ever was in reality. The narrator sees "the holy city Jerusalem coming down out of heaven from God." It has jeweled walls and pearl gates and streets of gold. The earliest renderings of heaven in Christian art depicted the "Heavenly Jerusalem": a mosaic from 440 CE on the walls of Santa Maria Maggiore in Rome shows a glittering blue-gold city, with the apostles, rendered as lambs, waiting outside the gates.

All of the churches in America's cities do good deeds in one way or another—homeless ministries and soup kitchens, economic renewal and education programs—and many of them use "New Jerusalem" language to inspire and motivate their volunteers. In 1993, Anthony Pilla, then the Roman Catholic bishop of Cleveland, launched "The Church in the City," which still gives grants for urban renewal projects. In a speech, Bishop Pilla made the link between a better Cleveland and heaven. The New Jerusalem, he said, "is a promise, a challenge and an invitation . . . to begin now to participate in the life of that heavenly city by practicing the mercy and justice that will make our earthly cities a reflection of that city which is to come . . . even as we wait for new heavens and a new earth, let us begin to build a new city of justice and peace." A Salt Lake City urban planner named Mike Brown, who is a member of the Church of Jesus Christ of Latter-day Saints, fantasized on his blog about the opportunities in heaven for a man with his skill set. "I dream of someday sitting in the back of a room and having the Lord Himself call on me to be part of the committee to design the New Jerusalem."

Most of us imagine that heaven is all these things—city, garden, banquet, home—and none of them, too. It is, we tell ourselves, beyond human understanding. Owen Gingerich is an astrophysicist at Harvard

University who spent his boyhood in a small Mennonite community in Iowa and continues, at age eighty, to believe firmly in a supernatural, creative God. Gingerich is an expert both in "the heavens"—that is, the sun, the moon, the stars, and the forces that make them move, as well as galaxies beyond the ones we can see with our telescopes—and in "heaven." He has participated in innumerable Faith versus Reason debates, debates he considers, on some level, fruitless. God is God, he would argue, and nature is nature. I visited him because I hoped he might have some idea about where heaven, physically, might be.

He does, but his idea would not help anyone seeking to find heaven in a spaceship. Gingerich is a small man with a full head of white hair, and on the day I met him he was wearing jeans and a casual blazer. He is known for being something of a jokester: when teaching Newton's third law of motion to undergraduates, he used to propel himself out of the lecture hall using a fire extinguisher. The afternoon I visited Gingerich followed a wet and stormy morning, and the professor's parrot-blue rain gear—jacket and pants—was draped over the chairs in his office. When I asked him, "Where is heaven?" he rose and, without explanation, walked over to a shoulder-high safe within his office. From it, he pulled out a small English almanac dated 1592—and illustrated with drawings of the universe as conceived by Copernicus.

Copernicus envisaged all the planets circling the sun. In one conceptual sweep, he overthrew all the Aristotelian conventions about the earth's place in the universe, and, by extension, the location of heaven. Earth (and its inhabitants) was no longer neatly contained beneath the sphere of the fixed stars, protected and governed by God in heaven. Copernicus described a boundless universe; God's home—now also out of bounds—had to be reimagined completely. "There was a lot of trauma with abandoning heaven as a nearby, physical place," Gingerich says. "People had to get over that."

Within the pages of the four-hundred-year-old pamphlet was a small, folded map, which Gingerich showed me, handling his precious book casually, as though it were the latest issue of *National Geographic*. There was the sun, encircled by Mercury, Venus, Earth, Mars, Jupiter, and Saturn—the only planets then discovered. Surrounding them all, marked by tiny stars extending outward to the edges of the page, was the infinite universe. Printed in the margins was this explanation by the English astronomer Thomas Digges: "This orbe of stares fixed infinitely up extendeth hit self in altitude sphericallye." It is here, beyond the stars and the orbiting planets, that heaven lies, "the very court of coelestiall angelles devoid of greefe and replenished with perfite endlesse joye the habitacle for the elect."

The more we know about the universe, the more we have to reimagine heaven, the "habitacle for the elect." We know for sure it is not in the sky, Gingerich tells me, nor is it anywhere in the known universe. For Gingerich, who by profession imagines places he cannot see, the out-of-place-ness of heaven presents no conceptual problem. Modern astrophysicists talk about multiverses, parallel universes beyond the spaces we know, governed by physical laws other than the ones we understand here. "These multiverse spaces are very much like heaven," he tells me, "something we can conceptualize but never observe." Heaven is simply . . . somewhere else. Its location is, as Gingerich says, "unanswerable"—but the question does not keep him up at night.

More challenging to Gingerich is the idea of submitting his individuality to the perfection of eternity, for humans are biochemical organisms that change from one decade to another—indeed from one millisecond to another. We grow, we learn, we remember, we forget, we deteriorate. We eat, we digest, we excrete. We make love, we sleep, we awake slightly different from what we were yesterday, our fingernails and our hair invisibly longer. The tendons that move my fingers over this keyboard wear out over time, the skin on my hands grows more delicate. Yet the movement of my fingers produces this book—eventually—and as I write I learn things I didn't know before. For Gingerich, the most difficult question about heaven is not where but how. How will the human organism, so defined by change, exist in eternity?

"Personally," says Gingerich, looking at me over his wire-rimmed glasses, "I say heaven is a great mystery. If I'm me, and I have an infinite amount of time, what will I do to stop from being bored? I imagine I will be learning Arabic and Sanskrit and learning it and forgetting it over and over again. I would like to see my mother again. But am I going to see my mother again as a thirty-year-old and me as an eighty-year-old? . . . What is it that constitutes *me*? How can it be preserved when we're part of an ever-changing stream? Are we going to be in a changeless state? But if we're learning, we're changing. The whole concept is so full of conundrums; one has to have hope in one type of continuity, but a continuity so inconceivably different from what we're in now that you get a headache from imagining it."

How much easier, says Gingerich, to imagine heaven as they did in the Middle Ages, a nearby physical place where God lived with saints and angels, arranged in neat rows, like a church choir. With our angel-studded Christmas cards and blue-sky children's books that show heaven as a pretty place "up there," Gingerich says, we are "still stuck—with one foot in the Middle Ages and the other foot on this modern view of space." I go home on the Amtrak train, wondering how on earth people might take comfort in a place they cannot know.

THE MIRACLE

D o Jews believe in heaven?" I can't begin to count the number of times I've been asked this question. Modern people—and especially Jews raised in the twentieth-century Reform tradition—are flummoxed by the notion of heaven, and the rabbis are less than helpful. "Jews believe it is this life that matters, not the next"—you hear this a lot. When my own congregational rabbi, Andy Bachman—who is Reform, a lover of baseball and political blogs, a father of three who picks his children up from school—talks about heaven, he refers to his own "conscious effort to teach people a reasonable Judaism." To the bereaved, he will talk about continuity from generation to generation. If a grieving person is receptive, he'll talk about souls ascending back to God, but he doesn't embroider—he doesn't speak of friendships in heaven, or family reunions, or continuing consciousness, or even selfhood. Rabbi Bachman doesn't talk about heaven's acreage or architecture; he certainly doesn't offer up the hope of bodily resurrection.

Judaism's vagueness on the matter of heaven is a problem for many of the Jews I know. An emphasis on this world may move them to honor their parents and to give to charity, but it's cold comfort when the time comes for those parents to die—or, worse to the point of being unimaginable, if a child should die. When you are facing the sheer cliff of your own mortality, the modern Jewish emphasis on this life rather than the next can feel brutally insufficient.

It may astonish readers to know, then, that it was Jews who invented our idea of heaven. They did not invent the idea of an afterlife, or the idea of heaven as the home of God—those ideas had been around for thousands of years, long before the Jews ever existed as a people. But the idea of heaven as we understand it—a place in the sky where the righteous go after death to live forever with God—that is a concept born to the Jews sometime during the second century before Christ. It was, if you will, a theological miracle. Similar ideas existed in other cultures around that time, and some Jews had even come to believe in a spirit world where their ancestors rested—and sometimes awoke— but the connection between "righteous" behavior, as the Bible puts it, and resurrection and eternal life was entirely new and almost entirely Jewish. Before the second century BCE, regular people, it was thought, didn't go to heaven. After the second century, some of them did. Later, Christians embraced the idea of heaven, amplified and embroidered it—and changed it in fundamental ways. Still later, Muslims made their Paradise both more coherent and more vivid than any Christian or Jewish version. But heaven, at its root, is a Jewish idea.

AFTERLIFE IN THE HEBREW BIBLE

Let us begin, then, with the Jewish patriarch Abraham, the man who received promises from God and who was willing to sacrifice his own son Isaac because God commanded him to do so. The Bible says nothing about what, if anything, Abraham believed about an afterlife. It says that his beloved wife, Sarah, died at the age of 127, and that Abraham purchased a burial cave called Machpelah for 400 shekels. He mourned for her, then buried her. When Abraham died, his sons Isaac and Ishmael, who had not been on speaking terms, came together to bury him in the same cave. Abraham died what people today would call "a good death," as the book of Genesis says, "in a good old age, an old man and full of years, and was gathered to his people."

But not, as far as the Bible indicates, to God. For the authors of the Torah, heaven was the home of God and his angels, but it was not the dwelling of humans or anything resembling humans. The Hebrew Bible is full of allusions to heaven, but all of them refer to the sky or the weather—such natural disasters as Noah's flood occur thanks to God's intervention on earth from on high. In the story of Lot, "the Lord rained on Sodom and Gomorra sulfur and fire from the Lord out of heaven." Heaven is the place from which God descends to give orders and to change people's lives. When God wants to tell Abraham that Hagar's son Ishmael will father nations, he speaks from heaven. When He commands Abraham not to slay Isaac, he sends down an angel from heaven.

Heaven is not just inaccessible. It is forbidden. Just before the Lord promises Jacob that he will inherit the land of Israel, Jacob lays his head down on a rock and has a dream in which he sees heaven's gate, an image that terrifies him. (In *Jacob's Dream*, a mid-twentieth-century painting by the Jewish painter Marc Chagall, the terror is gone; what Jacob dreams is something like the deep blue of a swimming pool, in whose depths people hide and float; on the ladder to heaven, angels twist and turn like circus performers. Jacob, the seer, wears a Mona Lisa smile.) The Bible is not so whimsical—it practically thunders as it describes Jacob's reaction to his heavenly vision: "He was afraid, and said, 'How awesome is this place! This is none other than the house of God, and this is the gate of heaven!" When people tried to build a tower that reached up to heaven—Babel—God did not allow it. He toppled the tower and scattered the people all over the earth.

When Jacob dies, he doesn't talk about going to heaven. All he wants is to be buried in his grandfather Abraham's cave. Jacob lives in Egypt, but he explicitly instructs his loved ones to carry his corpse back to Canaan. "I am about to be gathered to my people," Jacob says, "in the cave in the field at Machpelah, near Mamre, in the land of Canaan," where his grandparents are buried. Egypt is more than 200 miles from Canaan. Jacob's sons take the unusual step of mummifying their father's body before they carry him back to the homeland, where he is, according to his wishes, gathered to his people.

For the authors of the Torah, being gathered to your people was the best you could hope for after death. The ancient Israelites were buried in caves. These were not natural fissures, but man-made tombs, dug underground, usually with a central chamber and one or two adjacent rooms. When a person died, his or her corpse would be placed on a slab inside one of the chambers. When the next family member passed away, the old bones would sometimes be swept off the slab to make room for the new corpse. Generations of bones ringed the circumferences of these burial caves, intermingled and indistinguishable. Ancient Jews were literally gathered to their ancestors. Burial caves are still to be found everywhere in Israel, in orchards and beneath apartment buildings; bulldozers at construction sites often dig them up accidentally. Even today, those caves are full of bones.

Abraham and Jacob wanted their bones to be gathered together with those of their ancestors in the Jewish homeland, not elsewhere. This desire to be buried "at home" was then—and remains today— a critical theological and political point. Even a Jew who, like Jacob, spent his life, his career, his productive years abroad, wanted to spend eternity in the same cave as his ancestors. In America today, Jewish funeral homes with an Orthodox clientele offer Promised Land burials for about $5,000; clients of the Diaspora, who have spent their whole lives in Brooklyn, or St. Louis, or Atlanta, can join their forefathers after death in the dirt of the Promised Land. Moe Goldsman, a Los Angeles–based mortician, gave his wife a fat manila envelope for their tenth anniversary. Inside were deeds to funeral plots in Israel—in the town of Beit Shemesh, ten miles outside Jerusalem. "She was very upset," recalls Goldsman by phone, with a laugh. "Both of our families are buried here in Los Angeles. 'The kids aren't going to visit us,' she said." As a consolation, he also gave her a trip to Hawaii.

The Jewish claim that the bones of the patriarchs rest beneath the earth in Israel are—without some miraculous advance in DNA testing or the discovery of, say, a body buried with a magical staff or a multicolored coat—unprovable. Indeed, some recent Palestinian scholarship argues that without archaeological evidence of the patriarchs, Jewish claims to the land of Israel are empty. "So take the past, if you wish, to the antiquities market," writes the Palestinian nationalist poet Mahmoud Darwish. "Leave our country, / our land, our sea . . . everything, and leave / The memories of memory." The alleged site of Abraham's cave—long a tourist destination and a place venerated by Jews and Muslims alike—is in modern-day Hebron in the occupied West Bank and has been at the center of deadly disputes as recently as 1994, when the Israeli settler Baruch Goldstein opened fire on a group of Muslim worshippers there, killing twenty-nine of them.

The caves and the bones in them are so important because Jewish identity was rooted then—as it is now—in ancestry and not in an explicit promise (as, say, Roman Catholic and Islamic theology has it) of community togetherness in the afterlife. According to traditional interpretations of the Bible, the patriarchs had no heaven. What they had was generations of family and tribal affiliations—people in the past and in the future living under Abraham's covenant with God. ("As for me," God says in Genesis 17, "this is my covenant with you: You will be the ancestor of a multitude of nations. . . . I will give to you and to your offspring after you the land where you are now an alien, all the land of Canaan, for a perpetual holding.") A person's identity, in other words, was entirely connected to who his or her parents and grandparents were, and, after death, to the righteous behavior and community status of children and grandchildren. The worst thing for a Hebrew was to die childless. The prophet Isaiah describes seven women so desperate for children that they all appeal to one man, and offer to sleep with him, provide their own food, and make their own clothes, if only the man will "take away our disgrace!" The best a Hebrew could hope for was to rest for eternity among the bones of his people.

It would not, however, be strictly true to say that the biblical account offers no afterlife at all. There was a place called Sheol. The book of Psalms refers to this place, as do the books of Isaiah and Job. It was a dark and murky underworld where people had some sort of ghostlike existence. People whose lives had been unfulfilled—by displeasing God, or dying young or violently, or having no children—went to Sheol. It is cold, dark, mute, numb—entirely disconnected from God. In the Bible, nobody wants to go there. And from Sheol there's no return. Job mourns:

> As the cloud fades and vanishes,
> So those who go down to Sheol do not come up;
> They return no more to their houses,
> Nor do their places know them any more.

The Bible offers an inconclusive picture of life after death. You may be gathered to your ancestors. Or you may go down to Sheol, where everyone forgets all about you.

ANCESTOR WORSHIP

The biblical account of the afterlife is official teaching, handed down by the rabbis and codified in the Torah. It reflects, in other words, what the rabbis *want* people to think and practice—and perhaps even what many people did think and practice—but it is not an unbiased, historical account of what the people of the Bible actually believed. The Israelites of the biblical period were able to hold contradictory ideas in their heads about God and the afterlife. There was what the rabbis taught. And there was what they practiced, with their neighbors, at home.

Rachel Hallote is a tiny, birdlike woman with frizzy black hair that's graying at the temples and huge, hazel eyes. She is a dead ringer for her mother, the novelist Cynthia Ozick. An archaeologist by training and a professor at Purchase College, SUNY, she has spent decades excavating and studying the burial sites of the ancient Israelites, and her book *Death, Burial and Afterlife in the Biblical World* is an inspiration. Simply told, beautifully written (archaeologists don't usually write like novelists; Hallote does), it argues that the tombs of the dead Israelites tell an entirely different story from the stories of the Bible. She argues that while the Bible might insist on certain conventions concerning the afterlife—the righteous sleep like Abraham, others (explicably or inexplicably) go to Sheol—everyday practice was something else. Hallote believes that the ancient Hebrews were, like their neighbors in the land of Canaan (indeed, like other ancient or primitive cultures) engaged in ancestor worship. Their dead relatives *lived* somewhere, in another realm. They needed attention.

The archaeological record is convincing. In Hebrew burial caves of 2000 to 1500 BCE, Hallote and her colleagues have found jars and bottles for liquids and oils, small plates for food, weapons, and jewelry. I visited Hallote one rainy morning because I hoped she would clear up my confusion. The Hebrew Bible does not speak about a distinct afterlife. But if the Hebrews had no afterlife, why were they buried with all this stuff? Hallote's big eyes grew bigger. "Why are they leaving stuff? Because they believe the dead need it. . . . You've got to take care of your dead, because if you don't, they might wreak havoc on your life." The earliest Israelites believed dead spirits—good ones, bad ones—lived somewhere. It wasn't heaven, but it wasn't the peaceful sleep of Abraham and Jacob either. The idea that spirits existed in another, Sheol-like realm, that they communicated with the living, that the living could communicate with them and satisfy or displease them with daily actions—all these ideas existed in the Jewish context long before Jews ever started to talk about heaven.

Great books—*The Undertaking* by Thomas Lynch and *The American Way of Death* by Jessica Mitford—have been written about Americans' disconnection from death—about our habit of warehousing our elderly relatives until they die, then submitting them to chemical preservation and burying them in impenetrable lockboxes. In the ancient world, people were not so detached from their dead. The Hebrews lived in multigenerational family clusters, in small houses on top of their family burial caves. Within this tiny universe, generations would shift around in the rooms—the childbearing couple would sleep in a master bedroom; grandparents would sleep next door. And when the grandparents died, they would simply move into the cave beneath the house. (Hallote speculates that people who died in their prime would be buried farther away—in fields or farmland—where they could fend for themselves.) Like their neighbors the Canaanites, the ancient Hebrews did what they could to tend to and placate the spirits of their dead who lived beneath their floor. They fed them wine, they provided food, they gave them ointments. The Hebrews held banquets and parties in honor of their dead. They probably observed such tribal rites for thousands of years.

Hallote and others argue that the absence of afterlife in the Hebrew Bible was an effort by the religious authorities to suppress the cult of the dead among the Hebrews. The monotheism to which they were committed could allow no deviation. Wining and dining your dead, asking for their advice, holding feasts in their honor—all this looked dangerously like a violation of the First Commandment: I am the Lord your God, you shall have no other gods before me. One time-honored way to suppress unorthodox activity—in government, in the military, in religion—is simply to pretend it doesn't exist.

As Hallote and Segal, and indeed many other scholars, point out, the religious authorities who wrote the Bible knew that ancestor worship was part of Hebrew culture because they wrote strict laws forbidding it—and who writes laws against practices that don't exist? Deuteronomy, though chronologically fifth in the Torah, is thought to have been written first. It is, essentially, a list of dos and don'ts for the Jewish people, and among the don'ts are sorcery, magic—and the summoning of the dead. "No one shall be found among you . . . who practices divination, or is a soothsayer, or an augur, or a sorcerer, or one who casts spells, or who consults ghosts or spirits, or who seeks oracles from the dead," it says. In Leviticus, the authorities promise that anyone who raises the dead shall be cut off from God forever and "stoned with stones."

Those hoping to escape such punishment need look only at the story of King Saul for a deterrent. Saul, who historians believe lived around 1000 BCE, ordered all mediums and wizards expelled from his territories. Yet, faced with a swelling army of Philistines (an enemy tribe in Canaan), he became desperate and was driven to seek the counsel of his deceased mentor, Samuel. Saul knew he was violating the law, so he disguised himself in a cloak and went after dark to a sorceress to ask for her help in raising Samuel's spirit. Literature through the ages echoes the Bible: no good ever comes of talking to ghosts. In the underworld, Homer's great hero Odysseus sees his mother—who died from missing him—but he cannot touch her. Hamlet talks with his father's ghost, a conversation that leads to a stage full of corpses. When Samuel comes up from Sheol to talk to Samuel, he is cranky, like a child awakened from a nap. "Why have you disturbed me by bringing me up?" he asks. Then he tells Saul his destiny. He will be cut off from God. His sons will die in battle. The Philistines will win the war. The Bible could not be clearer: raise the dead and bad things will happen.

While reading Hallote's book at the New York University library one sunny afternoon, I found myself in a kind of waking dream. I began to imagine myself as one of the ancient Hebrews—as I am, a mother, a daughter, a sister, a wife. What if I had inherited a tradition, thousands of years old, of caring for and talking to my dead relatives? What if I prepared meals for all the people for whom I was responsible, both those living in my house and those buried beneath it? What if—God forbid—I kept my dead children as the Hebrews did, in jars under the house, where they remained under my protection? What if such rituals gave me comfort, continuity, a sense, however fragile, of control over the fortunes of my family?

Then, what if my rabbis told me that all this was forbidden? That these family customs violated God's laws? What would I do? How would I think about my dead? Wouldn't I want to imagine them in a place where someone else would be caring for them, with as much concern and attention as I could give them? Wouldn't I want assurances that they were safe, well fed, and cared for? If my dead spirits could no longer inhabit my house or my life, wouldn't I want to imagine someplace for them like heaven? I asked this question of Hallote: With their almost ruthless insistence on monotheistic orthodoxy, didn't the religious leaders pave the way for heaven? It's possible, Hallote answers, slowly. "The religion got rid of an aspect of religious life that was working. Certainly, by the sixth century BCE, you're not supposed to be communicating with your dead." By suppressing one kind of practice, the religious authorities created a hole, a space, in the hearts and minds of the Jewish people. Perhaps this need made them especially receptive to the new ideas they were about to encounter. Perhaps prohibitions against ancestor worship opened the way for heaven.

ZOROASTRIANISM

The Bible says that after Moses led his people out of slavery, after the Hebrews wandered for forty years in the desert, after God gave him the Ten Commandments, after Moses died without ever setting foot in the Promised Land, after Joshua led the Hebrews back to Canaan, fighting and burning his way northward, King David—the fierce, flawed, passionate poet—marched into Jerusalem with banners flying. David's men carried with them a box, covered with blue cloth, that contained the stone tablets God had handed down to Moses. David's son Solomon, following instructions from God, built a temple on Mount Moriah, the hill in Jerusalem where, generations before, Abraham had drawn his knife to sacrifice his son. The story was complete, the symbolism perfect. The Jews were back in the land of their ancestors, their monotheistic religion intact, their temple—which contained an inner sanctum for their Ark—a shrine to God's favor. Jerusalem became the center of the Jewish world, the Temple the center of Jewish worship, and for four hundred years, peace and prosperity reigned.

Then, in 586 BCE, the Babylonian king Nebuchadnezzar the Great stormed Jerusalem and drove many of the people who lived there—the elites and the skilled workers—into Babylon, roughly modern-day Iraq. The majority were left behind, too poor and uneducated to merit the king's attention. Nebuchadnezzar's armies smashed the Temple and the wall around the city, crushing the stones the Jews had so carefully quarried and carved. The Ark itself, containing the tablets from God, went missing. This loss remains at the center of the Jewish story.

What was the Jewish experience in Babylon? How did they practice their religion among strangers? More important, how—or rather, how much—did they adapt the religious beliefs of their new neighbors? For our purposes, these questions are crucial. Scholars believe that, after about a generation, the Jews lived in peace and prosperity in Babylon, too. The Persian king Cyrus conquered Babylon in 539, and after that time the Jews in exile lived as citizens in the great Persian Empire—territory that extended west to modern-day Turkey and east to India—contributing taxes and troops as needed. Persia was at that time the center of the world: its values were literate, enlightened, and tolerant. The Jews were free to practice their religion then, and they did so, worshipping at temples in Babylon and even on the Nile River island of Elephantine.

It was in Babylon, scholars say, that the Jews began to think of themselves as a people united by God, rather than a tribe connected to their turf and the day-to-day devotions at their Temple. For the first time, their identity transcended land and family. In exile, the Jews began to write down their codes, their laws, their stories. Far from their Temple, they began to create a religion that was "congregational," rather than merely local; their rituals and prayers transcended geography.

Here, though, is the important point: the Jewish exile coincided with the growing popularity of a new religion in the land that is now Iraq. This religion had originated about a thousand years earlier in Central Asia, in the mountains of northern Pakistan and Afghanistan, and was rooted in the teachings of a prophet named Zoroaster, or Zarathustra. It was a dualistic religion, which taught that all the good in the world came from the main deity, Ahura Mazda, who was associated with light, and that evil sprang from a demon named Agra Mainyu, associated with darkness. Zarathustra preached about an afterlife in which souls were judged according to their deeds on earth. At the judgment, each soul would walk across a razor-thin bridge that extended across a stinking pit; the good would pass safely across the bridge, while the evil fell in. And at the same time, he taught, the armies of light were at war in another realm with the armies of darkness. This war would bring about a cataclysm, a series of Messiah figures would come, and the earth would be refreshed, renewed, even perfected. Dead bodies would rise and be reunited with their souls. Everybody, in the end, would arrive in this Paradise, whose inhabitants would be, according to the Zoroastrian scripture called the Avesta, "non-aging, immortal, non-fading, forever living, forever prospering." The word *Paradise*, remember, has Persian roots.

Jamsheed Choksy, a professor of religious studies at Indiana University, believes that Zoroastrianism had a crucial impact on the development of Jewish (and eventually Christian and Muslim) ideas of heaven and hell. The Jews would have come across these ideas as they traveled through the empire on business, or as troops in the Persian army. The Israelites, he explains, even "had a representative at the royal court . . . [their rabbis] would have had theological debates with the magi [the Zoroastrian priests]. They probably had encounters and discussions." The book of Leviticus's fixation on purity, says Choksy, comes from the magi: "Clearly the magi and the rabbis are discussing what they should and shouldn't be doing." (And obviously the earliest Christians knew about the magi—for it was these same magi who made their pilgrimage to the manger in Bethlehem.)

But the Jews felt their monotheistic faith was threatened by the idea that good and evil came from two different places, and some scholars believe that the following verse in Isaiah (also written during the Exile) is a pointed rejection of Zoroastrian polytheism. "I form light and create darkness / I make weal and create woe / I the Lord do all these things." God creates the world, the prophet says, and everything in it: good and bad, light and dark. There are not two Gods: only One.

The story of heaven—that is, the story of how Western people began to think that they and their loved ones would go there—is bound up with the older story of Judaism caught between cultural currents, under pressure to assimilate into the mainstream on the one hand and under equal but opposite pressure to retain its unique theology and identity on the other. I have marveled, in doing the research for this book, at how enduring this theme is: the Jews, having weathered one historical cataclysm after another, continue to live caught between the extremes of liberalism and orthodoxy. In the neighborhood where I live in Brooklyn, the children of Jews like me—married to a non-Jew, celebrating and observing an amalgamation of traditions—play on the tire swing in the playground with the children of the ultra-Orthodox, the boys with their heads covered and their tzitzit showing where their dress shirts meet their loose-fitting trousers. We, the parents, smile warily at each other, reenacting the perennial tensions in our religion.

Our story now moves back to Jerusalem, where the idea of heaven was forged out of such tensions. In 200 BCE Jerusalem wasn't much of a place. It had not yet recovered from Nebuchadnezzar's siege. It was a ruin, a place so reduced that no one in the Ancient Near East would have said it mattered, or even called it a city. Only seven or eight thousand people lived there, most on a single hill in plain view of where their Temple used to be. These survivors had rebuilt the Temple, probably from the stones of the first, but it was smaller, and shabbier, than the original. Every time they sacrificed a goat or a lamb at its altar, the Jews must have been reminded of everything they'd lost: their Ark, their Temple, and their identity as a community that existed to serve God at that Temple. The remaining inhabitants of Jerusalem thought of themselves as a humiliated and desecrated people.

But the city was beginning to change. The Jews of the Diaspora had grown prosperous. Trade routes had opened between Jerusalem and Babylon and points east, and between Jerusalem and Egypt and points south. These Hebrew merchants and tradespeople spoke foreign languages. They stored precious stuff—vases, plates, leather goods, coins, food—in the Temple. They knew something of the wider world. They could read and write: they were codifying the stories and laws that would become the Torah. Still, despite all these innovations, they did not—exactly—believe in heaven.

Archaeology tells us the tribe that became the Jews had participated in something like the primitive ancestor worship of their Canaanite neighbors in the centuries before the Babylonian exile. They probably also knew about Egyptian death and burial rites, because Jacob was mummified. Perhaps they also knew that the souls of the pharaohs were thought to fly up to the stars to live as gods, and, as we have seen, the Jews of Babylon knew something about the teachings of Zoroaster and his vision of a final judgment where the good go up and the bad go down. Still, the idea of souls ascending to live with God in the sky was not yet born to the residents of Jerusalem.

Like a cheesy magician, I have left a card up my sleeve. I've mentioned the afterlife alternatives given to Jews by the authors of the Torah: either eternal rest among the bones of your ancestors, or Sheol, a shady place where no one wanted to go. The Torah metes out punishment for anyone who raises and talks to their dead. And that, I seemed to say, was that.

Except that *wasn't* that—not quite. Two characters in the Hebrew Bible do ascend to someplace like heaven, and there they live with God, though it's not clear how—or why—they earn this honor. The first one is Enoch. The Bible gives almost no personal details about Enoch: his name is mentioned in Genesis, in a list of begats; he is a son of Jared, and the father of Methuselah. Yet, in a single sentence, the Bible imputes to him an extraordinary destiny: "And Enoch walked with God; then he was no more, for God took him." He no longer existed on earth, in other words, he existed with God. No mention is made of how he died—or indeed, *if* he died—or what he did in life or whether or not he was worthy. He simply went to live with God. Genesis is thought by most scholars to have been written and finalized toward the end of the fifth century BCE, or perhaps somewhat earlier. In the millennia that followed, Jewish, Christian, and Muslim stories about heaven often included an appearance by a personage named Enoch; some versions were even written by "Enoch" himself. The most influential of these visionary tales were gathered together as the book of Enoch—a compilation of apocalyptic works written between the second century BCE and the second century CE by a succession of anonymous Jewish and early Christian scribes.

The other character who ascended to something like heaven is Elijah, the prophet found in the Hebrew Bible, in the books of Kings, the same prophet whom modern Jews welcome at their Passover tables each spring. Elijah served as the mouthpiece of God. He praised God's glory, he foretold His fury, he warned Jews about the punishments that would befall them were they to worship anything or anyone else besides the God of Israel. Elijah, according to the Bible, was just walking along with his disciple Elisha when he was "taken up": "A chariot of fire and horses of fire" appeared, "and Elijah ascended in a whirlwind into heaven." In my research for this book scholars have—repeatedly—warned me against overgeneral declarations. The Hebrew Bible makes no mention of people going to heaven—except in the two instances where it does. Some scholars say that by the time of the Exile, some Jews, some of the time, believed that some special people went to heaven.

HELLENISM

And I've also kept another card up my sleeve. Despite all the influences that neighboring religious practices may have had on the way the Jews imagined heaven, most notably the fusion of Zoroastrianism with traditional Jewish beliefs, no influence was more important to the invention of heaven than that of the Greeks.

In 333 BCE, two hundred–odd years after the Babylonian exile, Jerusalem had been gobbled up in Alexander the Great's sweep southward. The tiny city itself did not matter—scholars debate whether Alexander ever went there—but it was, like so many other small cities, a pawn in the political power struggles that emerged after Alexander's

death. It became part of the great Hellenistic Empire, a region that absorbed most of what was once the Persian Empire: extending as far east
as the Himalayas and as far west as modern-day Turkey. In Israel today,
a tourist can see Greek ruins, the remnants of temples dedicated to
Greek gods. In the third century BCE, the language of commerce and
government was Greek, education systems were Greek, and taxes were
paid to provincial governors with allegiances to the Greek emperors.
Jerusalem was not an official city-state, and the Greeks allowed the
Jews more or less to govern themselves, but the culture around them
was pervasively Greek. Like Jews or Muslims in America today, whose
discomfort with such mainstream, quasi-religious symbols as Christmas trees and Easter eggs is mixed with a longing to embrace them,
the Jews of the third century before Christ lived within a dominant
culture so powerful it would have been impossible to evade.

As any eighth-grader knows, the Greeks had a pantheon of gods,
but these gods would have mattered very little in any practical or day-
to-day sense to the Jews in Jerusalem. Before around 200 BCE, the
Jews, as a self-governing people—the priests of their Temple were their
highest authorities—were not required to participate in what amounted
to the pagan state religion. But the worldview of the Greeks was un-
avoidable, and it changed forever not only the way they saw themselves
but their visions of the hereafter.

The Greeks, radically unlike their Jewish neighbors who were tied
to generations of family history, emphasized the accomplishments of
individuals. Their well-known love of sports and games reflected a
more general love of individual competition. Their schools taught
reading, writing, history, literature, and mathematics, and young boys
learned that their individual contributions to society mattered. Residents of the empire who were citizens (not the Jews) could vote—and
many of those who couldn't aspired to citizenship and held democracy

as an ideal nonetheless. Alexander's own favorite sculptor was Lysippos, who made the great conqueror look like a varsity athlete, hair sweaty and matted, brow furrowed from exertion, rather than the customary bland, anonymous leader-god.

This worldview presented the Jews with an alternative way to see themselves. The Jews who had gone into exile had already begun to grapple with their identity specifically *as* Jews, since they'd long been separated from their ties to the land. The influence of the Greeks released the Jews, at least to some extent, from their tribal and generational chains.

Even more critical: the Greeks believed in the soul, the life force within each person, which was also the seat of individual personality and agency. It resided, the Greeks believed, inside the human head. They were convinced—very broadly speaking—that the spirit and the body were joined in life and that after death they separated. The body was dirty and corrupt, the source of all lower human activity— sex, eating, childbirth, sickness. While it was left in the ground to decay, the soul ascended, free of constraints, to live—if it was worthy— with the gods. In Plato's *Phaedo*, Socrates says that the soul "departs to the invisible world, to the divine and immortal and rational: arriving there, she lives in bliss and is released from the error and folly of men, their fears and wild passions and all other human ills, and forever dwells . . . in company with the gods." The soul, it would seem, has a mind of her own.

For Plato, the highest human achievement was wisdom. In life, he taught, men should try to make their bodily desires submit to their intellect, their soul; in death, the captive soul would be set free. And the more wisdom a person accumulated on earth, the higher the soul would ascend toward God. (Plato believed in a creator god, which he called the Demiurge, as well as in a constellation of other gods, such as stars, which were perfect and eternal. Toward the debauched antics of the Greek pantheon—Athena, Zeus, and so forth—Plato had only disdain.) Those with insufficient wisdom would be sent back to the world for another try—Plato also believed in reincarnation. Humans were made of two separate parts: mortal body and immortal soul. "There is no question, hands down," Segal told me, "that the Greeks were the most important influence on what we think about heaven."

I like to think of these ideas as floating around in the atmosphere of Jerusalem, looking for a home. There was the nearly universal idea that God, or the gods, lived up in the sky. There was the Canaanite idea that the beloved dead lived on, in some way, needing attention, gifts, and sustenance. There was the Egyptian idea that kings ascended to the sky after death to live with the gods—and the biblical idea that in at least two cases, special people could, too. There was the Zoroastrian idea of separate afterworlds for the good and the evil, the final destinations to be determined after the return of saviors to earth and a last judgment. And—perhaps most important—Greeks had the idea of an eternal soul that ascended to live with God under the right circumstances. The Jews would need a cultural and historical big bang to fuse them together.

THE BIG BANG

Around the year 200 BCE, tensions began to emerge in Jerusalem over the Jews' relations with their Hellenistic overlords. The question, if you boil it down, was the same question that would continue to confront the Jews for thousands of years to the present day. How much should they participate in mainstream culture? And to what extent would such participation compromise their unique beliefs and religious practices? They had faced this question with regard to their

Canaanite neighbors when they were still a country people and Moses, coming down from Mount Sinai, thundered at them not to worship idols as the Canaanites did.

The inhabitants of Jerusalem imagined at least two different resolutions to their conflict. The priests and the nouveaux riches favored aligning themselves with the Greeks and transforming Jerusalem from a ghetto backwater into a recognized city-state. Such an alliance would bring commerce and wealth to the city; it would give Jews a voice in the government and Jewish children the sophisticated education for which the Greeks were known, and they would become athlete-scholars, democrats, philosophers. The poorer residents, however, who had scraped by on their little hill for centuries while their countrymen lived in affluent exile, wished for no such alliance. They were suspicious of the dominant Greek culture; the Greek pantheon would have been repugnant to them, the pagan customs of the Greeks—naked athletic competitions, for example—would have been abhorrent. For four hundred years they had lived in full view of Mount Moriah, reminded constantly both of Abraham's covenant with God and of the destruction of their Temple. They were content to remain small, insular, and separate.

The priests and the wealthy won out, of course. In 175 BCE, Jerusalem's high priest, who bore the Greek name Jason, paid the Syrian king Antiochus IV, who held Jerusalem in his control, for permission to build a gymnasium in the center of the city. The gymnasium polarized the Jews. In any Greek city, the gymnasium was the ideological center. It taught the Hellenistic values of competition, individualism, democracy, and learning. Its colonnaded halls were adorned with statues of Greek gods; in its plaza, naked young men learned sports and games. No one knows exactly where the gymnasium in Jerusalem was; its remains have not been found. Professor Hanan Eshel, former head of the Department of Land of Israel Studies and Archaeology at Bar-Ilan University, believes the gymnasium was built on a hill across from the Temple, where the city's rich had begun also to build large houses. When pious Jews took their animals to the Temple for sacrifice, they could see the sons of the priests, training for athletic contests, naked as the day they were born.

The books of the Maccabees—Jewish texts not found in the Torah—describe the goings-on at the gymnasium with horror. Young Jewish men wore wide-brimmed hats, it complains, in honor of Hermes, the fleet-footed god of the Greek pantheon. Rich Jewish boys, trying to fit in, endeavored to reverse their circumcisions. The author of one of these histories reserves special contempt for these young men: to hide circumcision is to abandon the covenant with God, he says. These boys "joined with the Gentiles and sold themselves to do evil."

The story of the Jewish revolt against the oppressive King Antiochus—which Jews know today as the story of Chanukah—is also the story of heaven. After Jason built the gymnasium, power struggles erupted among the other priests. One rival, named Menelaus, offered Antiochus IV a cash payment for Jason's job and Antiochus, needing money after decades of waging war against the Romans, accepted. Menelaus paid the king with treasure from the Temple's stores, and the city's poorer Jews—who would have been able to watch their precious belongings being hauled out of storerooms and onto carts bound for Syria—went crazy. For years, the Jews of Jerusalem fought each other, priest against priest, neighbor against neighbor, brother against brother.

Finally Antiochus had enough. In 167 CE he sent an occupying army into the city and set forth a series of decrees. Jews could not observe the Sabbath. They could not circumcise their sons. They were to sacrifice swine—forbidden by their Torah—on altars he chose, to gods he designated. He installed in the Temple what the books of the Maccabees call "a desolating sacrilege": a statue of Zeus. Antiochus's act was an unspeakable violation.

According to the books of the Maccabees, a poor and pious boy named Judas Maccabeus "got away to the wilderness and kept himself and his companions alive in the mountains as wild animals do." He began raising a guerrilla army until he had six thousand men. He rode into Jerusalem and managed to vanquish Antiochus's soldiers. He cleared the Temple of its abomination and rededicated it to the One God of Abraham. The meager quantity of oil in the lamp he lit lasted eight days and eight nights, and "the talk of his valor spread everywhere."

Somewhere on the sidelines, watching these bloody events unfold, sat a man whom I will call Daniel. As he watched, he wrote prophecy, predictions of events to occur in the future, and the book of Daniel contains the first explicit reference in Jewish Scripture to anything like what we know now as heaven. Daniel was apparently a teacher and a sage. At first, Daniel advised the Jews to ignore the fighting in the streets: sit tight and pray to God. But once the streets of his city were full of Antiochus's soldiers, he changed his mind. Perhaps he fled to a cave in the hills, where he wrote the words he hoped would inspire his people to passive resistance. Do not fight, Daniel said—but do not capitulate. Martyrs, he promised, would be rewarded for their faithfulness in a special way, and the words he wrote would change forever the way people imagined the future of their immortal souls. "Many of those who sleep in the dust of the earth shall awake, some to everlasting life and some to shame and everlasting contempt. Those who are wise shall shine like the brightness of the sky, and those who lead many to righteousness, like the stars forever and ever." With this verse, Daniel gave us heaven.

John Collins is Yale's Holmes Professor of Old Testament Criticism
and Interpretation, a scholar so revered and accessible that graduate
students wait in line outside his door like acolytes. An Irishman who
fell in love with ancient languages when he was in high school, Collins is an authority on the book of Daniel. I made a pilgrimage to New
Haven because I wanted Collins to help me answer a question: Why
did Daniel offer the prospect of heaven at this particular moment?
Earlier Jews, like Abraham, had wanted nothing more than to sleep
in caves with their ancestors. How and why did the cultural, religious,
and historical threads entwine at this moment—165 BC—to give us
this miraculous idea?

I should have known better. It was like asking an evolutionary
biologist to explain why the first creature slithered out of the slime
to live on land and breathe air. The smart scientist would say, "Who
knows?" and Collins is a smart scientist. Who was Daniel? Collins
shrugs, smiles, and lifts the palms of his hands upward. As he points
out, the ancient Israelites left behind few records. Daniel was a wise
man, a teacher. He could read and write. He spoke Hebrew and Aramaic. He knew the Bible stories and other Jewish narratives—including
the earliest passages of the book of Enoch—that have not become
part of the official canon. Collins interprets the book of Daniel as a
chronology of current events posing as prophecy. In other words, the
narrator puts himself in the past, as a person able to "see" what is happening in the future: the persecutions of Antiochus and the bloody
rebellion.

But why was Daniel promising heaven just then? The Jews had been
through tough times before—notably during the Babylonian exile—
and no one had talked about heaven then. It was enough for them
to die and be buried together. But by the time of the Maccabees, the
Jews had dispersed enough that often grandfathers, fathers, and sons

no longer lived in the same cities. "Being gathered to your ancestors" was not as simple and inevitable as it had been during the time of the patriarchs. And thanks to the influence of Hellenism, success for Jews had become as much a matter of individual accomplishments redounding to their own personal honor as to the honor of the family or a tribe.

Heaven, as the book of Daniel tells it, is a reward for an individual. Indeed, it is specifically withheld from those who do not follow the prophet's urgent call, who are not, in his judgment, "wise," and who fail to "lead many to righteousness." Being satisfied with sleeping forever in a cave worked, Collins says, "when the emphasis was on community and the family. The afterlife was your children and your name." For the author of Daniel, though, "the goal in life is to hang out with the angels. Permanently." At first, Collins adds, Daniel's auditors must have thought he had lost his mind. Nothing like this had ever been an acceptable part of Jewish thinking before. But the idea had been born, and it caught on fast. Within a hundred years after Daniel wrote, Collins says, many of the Jews of Jerusalem would have said they believed in heaven.

And then Jesus came along and changed everything.

THREE

THE KINGDOM IS NEAR

We recently had a funeral for our pet fish, Louis. He was a blue tetra, and from blue we got blues and from blues we got Louis, as in Armstrong. In the morning we noticed him drifting around in the tank on his side, unable to motor through the currents created by the bubbler. Then we found him whirling in the water in great, nauseating circles. Finally, around dinnertime, we found his little corpse clinging to the tank's filter. We dropped him in the toilet, we lit a candle, and I said, "Louis, you were a good fish and now we commend you to heaven." Then Josephine said, "Louis, I hope you are happy in heaven." And we flushed. It seemed to make Josephine feel better that Louis went straight to heaven, though in truth, she was less rattled by Louis's demise than either of her parents. We felt our failure strongly.

We comfort our children—and ourselves—by telling them that when we die we go to heaven. In Maria Shriver's 1999 children's book *What's Heaven?* a mother explains the afterlife to an inquisitive young daughter after the girl's great-grandmother dies. The way the book tells it, the great-grandmother's soul "has already been taken up to heaven," even before the funeral. By the time the body is in the ground, the spirit is happy with God. It's simple, the mom says. "You go to Heaven when your life here on earth is over."

Yet in truth, the question of *when* you go to heaven—especially among Christians—is extremely complicated. It's a theological brain twister, bound up in such seemingly unrelated issues as what you believe about Jesus, about history, and about the role of humans in enacting God's will. Most people believe in the Maria Shriver version of things. According to a 1997 *Time* poll, 61 percent of Americans believe they'll "go directly to heaven" when they die. But the orthodox monotheisms actually teach that the trip to the next world is a two-part journey. You go somewhere when you die. And then you go to *another* place at the end of the world. (The latter is what Jews call "*olam ha-ba*," the world to come, what Christians call "the new heaven and the new earth," and what Muslims call "Paradise.") In modern America, we usually call this whole series of events "heaven." Some people also believe that heaven is a place—or an occurrence—on earth, which reveals itself in profound and unexpected glimmers. These concepts are not necessarily mutually exclusive. Throughout the centuries, people have answered the question "When do you go to heaven?" in any number of contradictory, even illogical ways. While Jews and Muslims have their versions of these debates, the confusions that have filtered into our contemporary world are inherited mostly from the Christian story—and it is that story that I will primarily tell in this chapter.

WHAT JESUS MEANT

Jesus said a lot about heaven, much of it hard to understand. In the Gospel of Matthew alone, he talks about the Kingdom of Heaven in terms of a farmer sowing a field, a mustard seed, the yeast in dough, a fishing net, a merchant looking for pearls, a king settling accounts, a king who has prepared a wedding banquet for his son, a landowner who hires men for his vineyard, and ten virgins. With these parables, Jesus seems to be saying that heaven is a place both known and unknown, like the world but unlike it—a place of love and justice, big enough to accommodate all the souls in the world but open only to some. He is unambiguous on the question of the poor: it is easier for them to gain heaven than the rich, he says—who will find it harder to get in than a camel trying to fit through the eye of a needle. He advises followers not to hoard wealth on earth. "Store up for yourselves treasures in heaven," he says, "where neither moth nor rust consumes." He adds that to inherit heaven, people must become innocent, like little children. Righteous acts were important to Jesus—he was a Jew, after all, and had inherited a code of law that applied to every aspect of life—but humility was, too. "Beware of practicing your piety before others in order to be seen by them," he says in the Gospel of Matthew. Heaven is home for the righteous, but not for those who display their righteousness to impress other people.

On the question of "when you go to heaven," however, Jesus was far from definitive. In the Gospel of Mark, probably written within fifty years of Jesus's death, the Christian Lord says, "The time is fulfilled, and the kingdom of God has come near." For centuries, scholars have acrimoniously debated what that meant. N. T. Wright, for example, says that Jesus's announcement of the Kingdom had nothing to do with what happens when you die. He was posing a question: "What would it look like if God was running this show?" But the historian Albert Schweitzer argued the opposite in his 1901 book *The Mystery of the Kingdom of God*. Jesus was talking about "a final cosmic catastrophe," he wrote. When he said the Kingdom was near, he meant that life on earth was about to become unrecognizable.

Before diving into that debate, though, a word about history, and the way Westerners understand it. *Eschatology* is a theological word,

which means "the study of the end of the world." All three of the
Western monotheisms are eschatological. Jews, Christians, and Muslims believe that history is linear and progressive. It moves forward,
never back. (Buddhists, by contrast, believe that history is cyclical: the
universe comes into being, evolves, fades, and disappears again and
again.) In their most orthodox forms, all three monotheisms hold that
God exists outside history, intervenes in human events, and at the end
of time will judge mankind and create for the righteous a new world
(*olam ha-ba*) that is eternal and perfect—an actual, physical dwelling
place. *Apocalypticism*, another scholarly word, is a radical form of eschatology. Apocalyptic literature or scripture predicts God's imminent
intervention, perhaps at any minute, to bring about a cataclysmic end
and ultimate perfection for the righteous.

If you use a broad measure, you can sort believers—in all three
monotheisms—into right and left camps according to their eschatology.
Progressive believers tend to put eschatological expectations on the
back burner, content to deal in the here and now and leave last things
to come as they may. Fundamentalists tend to have a more apocalyptic view, a certainty that humankind is entering the final act—and
often a sense of their own divinely ordained role in that end. The
American version of fundamentalism is familiar enough, though rich
in its variety—ranging from members of the Church of Jesus Christ of
Latter-day Saints, who store cans of food and bags of rice in their basements in preparation for any eventuality, to members of the Jehovah's
Witnesses, who believe that all the righteous will soon be walking
around on a renewed planet, just like this one but better in every way
(while the super-righteous ascend to heaven to live with God). Former
Alaska governor Sarah Palin is this kind of fundamentalist. Video
clips that circulated during her vice presidential run in 2008 showed
Palin at her Pentecostal church, affirming her conviction that the war

in Iraq was a "task from God." While blue-state Americans regarded these remarks as evidence of Palin's unsuitability for office, other Pentecostals and scholars of Pentecostalism promised me that this rhetoric was simply boilerplate in such conservative churches. "There is absolutely nothing about this that I haven't heard a million times," said Grant Wacker, professor of Christian history at Duke.

Progressive Jews don't think too much about the end of the world, but ultra-Orthodox ones do, and since some believe their Messiah has come in the form of Rabbi Menachem Schneerson, certain Lubavitcher Jews are expecting the end any day. In the (Sunni and Shia) Muslim world, scholars say apocalyptic expectation is on the rise. "The Hour is coming," reads the Qur'an, "I am about to reveal it." Iranian president Mahmoud Ahmadinejad, for example, belongs to the Twelvers—a subset of Shiites who believe that Muhammad al-Mahdi, born in 869, the twelfth leader to succeed the Prophet Muhammad, did not die but was hidden away and will return soon as the Mahdi, or Messiah, to save mankind. According to this theology, which Ahmadinejad employs to great political effect, the destruction of Israel precedes the cataclysm that destroys the world and brings about perfection. In his September 2006 speech to the United Nations, Ahmadinejad referred directly to this belief when he spoke of "the real savior who has been promised to all peoples and who will establish justice, peace, and brotherhood on the planet." The State of Israel is similarly important to apocalypse-minded Christians who believe that Jesus will only come again when all Jews have gathered in the Holy Land. It is for this reason that conservative evangelical ministers like John Hagee—with whom Senator John McCain briefly allied himself during his 2008 presidential bid—see the protection of Israel as a theological imperative. The war over the Holy Land is, and has always been, literally, a conflagration about the end of the world.

Each monotheism has always included both eschatological hard-liners and those who are happy to let history unfold as it will—and everyone in between.

Where, then, do the teachings of Jesus belong on this spectrum? In America, both sides claim their view as his own. The Jesus Seminar of the 1980s—an effort to discern which of Jesus's sayings really originated with him—holds that Jesus was a rabbi, a teacher, a sage, even a mystic. He was perhaps divine, but in any case an advocate for social justice, for a radical overthrow of the existing hierarchical order, and for a redefinition of human relationships to extend beyond family to a community of the faithful. The Kingdom of God, according to these scholars, is two things: the ultimate Kingdom at the end of the world, and—more important—the community of believers on earth, endeavoring through their daily actions to mimic or replicate that heavenly abode. When Jesus said, "The Kingdom is near," he meant, literally, now. The movement he led, which his disciples and the apostle Paul carried on after his death and resurrection, was—and is—an explicit and conscious approximation of godliness on earth. "The Kingdom of God means . . . the ruling of God in our hearts," explains the *Catholic Encyclopedia*.

John Dominic Crossan, who codirected the Jesus Seminar, insists in his winning Irish accent that Jesus had no preoccupation with the hereafter, and he points to the Lord's Prayer, which Jesus himself may have said, as proof: "Your Kingdom come. / Your will be done, / on earth as it is in heaven." "You don't get anything about *going* to heaven in there," he says. "It's taken for granted that heaven is the dwelling place of God—there's not a single thing about life after death." Does he believe in heaven? Crossan says that, for himself, life on earth has been "absolutely, incredibly enough. . . . If you say, 'Do you want to go to heaven?' I'd like to know the options and conditions before I say yes."

Other scholars—most notably Schweitzer—look at the historical Jesus in the light of his Jewish context and draw a different conclusion. Most Jews of the first century took for granted a kind of watered-down apocalypticism, says the scholar E. P. Sanders, who retired from Duke University in 2005. They believed that God reigned in heaven and that at the end of time he would, as Sanders puts it, "govern everything perfectly." This scholarship makes Jesus a conceptual heir to Daniel, as well as to Ezekiel, Isaiah, Enoch, and other Jews who expected God to intervene and set things right *any day*. Jesus was a cousin of John the Baptist, thought by some to have been influenced by the Essenes—the ascetics who lived in monastic communities throughout what is now Israel waiting for the world to end.

When Jesus said, "The Kingdom is near," these scholars say, he was not talking about a kingdom in our hearts or in our communities but about a literal cataclysm—an explosion, a fire, a flood—that would bring about the end and usher in paradise in one gigantic and divinely ordained event. Jesus had an awareness, these scholars say, both of the imminent end of the world and of his own role in bringing it about. In the Gospels, he asks the disciples who he is and Peter replies, "You are the Messiah." Hush, says Jesus. "He sternly ordered them not to tell anyone about him." The disciples, who were also Jewish and familiar with apocalyptic prophecy, may even have expected that the death of their Lord would trigger the events that would usher in the end of the world.

But within a generation, the earliest Christians had a big problem. If Jesus had indeed been predicting the end of the world—if when he said, "The Kingdom of heaven has come near," he meant, "the sky is about to fall"—and the world didn't end, then Jesus was wrong. And if Jesus was wrong, how could he be God? The failure of history to fulfill the prophecy of Jesus and his disciples within a generation triggered centuries of speculation concerning the end and the exact order of events. The apostle Paul, in order to quell these concerns, advised believers to sit tight and stay faithful, for the end would surely come—it was just a matter of when. "For you yourselves know very well," he says in his first letter to the Thessalonians, "that the day of the Lord will come like a thief in the night. When they say, 'There is peace and security,' then sudden destruction will come upon them, as labor pains come upon a pregnant woman, and there will be no escape!"

Over time, says Sanders, the focus shifted—away from an immediate "when" and toward a theology that would work in *this* world. The Church Fathers, the Christian thinkers and writers in the first centuries of the common era whose work formed the foundations of the Roman Catholic Church, and eventually much of Christian theology, imagined solutions to the delayed apocalypse in numerous ways. They developed a vague "in between" phase, which by the Middle Ages had flourished into a full-blown purgatory. You died. There occurred an individual mini-judgment that sent your soul, or spirit . . . somewhere. Maybe—if you were a martyr or otherwise extremely righteous—it was heaven, where you would reside with God. Maybe it was some kind of passive waiting state, conscious or otherwise. (Irenaeus, the second-century Bishop of Lyon, favored this option, writing that souls "shall go away into the invisible place allotted to them by God, and there remain until the resurrection, awaiting that event.") Or maybe it was a period of active cleansing or punishment meant to prepare your soul for meeting God. Augustine, who was the fourth-century Bishop of Hippo, which is in modern-day Algeria, belonged to this last group. He believed some souls would have to endure intermediate purification by fire before they went to heaven. "This fire," he said, "will be more terrible than anything that a man can suffer in this life."

Ingeniously, certain Church Fathers also collapsed time. Once a soul *did* enter heaven—whether at death, after purgation, or at the end of the world—chronological time would cease to have meaning. The question of "what's next" thus magically disappeared. Eternity "is a never-ending present," Augustine wrote. "You are at once before all past time and after all future time. . . . Your years are completely present to you all at once, because they are at a permanent standstill."

All three monotheistic religions agree on the finale in its broad outlines. At the end of the world, the living and the dead are raised and judged by God. The wicked are sent to hell, or something like it. The righteous are ushered—with their bodies or something like their bodies—into paradise. And there, in the traditional Christian view, is the heavenly city and the garden: its walls sparkle with sapphire and onyx. The righteous enter the city through gates of pearl, and behold rivers flowing with water. In this new Eden grows the tree of life, heavy with fruit. There is no sun, and no moon, for everything is illuminated by the light of God. All the faithful are invited to partake of these glories. As the book of Revelation says, "And let everyone who hears say, 'Come.' "

Anne Graham Lotz is the second of the evangelist Billy Graham's five children. At sixty-one, she is a tall, storklike woman with iron-gray hair in a severe bob. The planes of her face are sharp, recalling her father's handsome, hawklike visage, and she has her father's voice, a Southern singer's honeyed rasp. Yet her manner is warm and direct, with a just-us-girls appeal. I like her. Through her I've met people who are now my friends, and I like that she—like so many ambitious women—clearly struggles with how to reconcile her ambitions with her obligations to her family. Despite her complicated history with her parents and her siblings (all of whom have flourished and suffered from having a father so famous, and so absent), she speaks of her upbringing with affection. Lotz's brother Franklin is the official heir to their father's legacy, the head of the Billy Graham Evangelistic Association. But Lotz has inherited her father's gift for preaching. She has her own nonprofit, called AnGeL Ministries, which is an evangelizing organization; she writes books and she preaches at revivals all over the world.

Lotz believes the return of Jesus (and so the apocalypse) will come in her lifetime—which puts her in good company. A third of America's white evangelicals, about twenty million people, believe they will live to see the end of the world. This is not a symbol or a metaphor, but a literal, historical event about to unfold. Lotz believes that the Bible, the inerrant word of God, contains the map or code revealing the wheres and whens. Recent disasters—9/11, Hurricane Katrina, the Indonesian tsunami that killed over 228,000 people in 2004, and the earthquake in China's Sichuan Province that killed an estimated 70,000 in 2008, war abroad, and the meltdown of global markets—all are the fulfillment of biblical prophecy, preludes to a final, cataclysmic battle between good and evil. "These are like alarms going off," she says. In the end, Jesus will return to vanquish the devil and reign in peace for a thousand years. Then God will raise the dead and judge them. At that time, heaven and earth will merge, and in that fusion the world, and everyone who has been saved through Jesus from destruction and hellfire, will be perfect.

Lotz says her sense of urgency separates her from her parents' generation. She doesn't recall her mother and father talking together with their friends about the return of Jesus in the same way. Lotz travels and teaches all over the world: to Korea, and Eastern Europe, and the African continent. "Everywhere I go," she says, "people are expecting the return of the Lord."

If Jesus does return to earth before her natural death, Lotz believes she and every other born-again Christian will ascend to heaven in an event called the Rapture, a sudden whoosh skyward—similar to what happened to the prophet Elijah who was "taken up" by God. The Bible does not use the word *Rapture*, and the concept has become widespread only in America over the past hundred years—although it has roots here as old as the Puritans. It's based on a verse in 1 Thessalonians. At the Second Coming, Paul says, the dead will be raised first. "Then we who are alive, who are left, will be caught up in the clouds together with them to meet the Lord in the air; and so we will be with the Lord for ever." The best-selling Left Behind series of novels by the evangelist Tim LaHaye and his coauthor Jerry Jenkins opens with the Rapture. Little puddles of clothes and jewelry are scattered over the earth, in armchairs and on airplane seats, their human owners suddenly no longer in need of them. The rest of humanity—including the book's hero, Rayford Steele—is left behind to duke it out with the devil. All told, the series has sold sixty-five million copies.

If, on the other hand, Lotz dies before Jesus returns, she believes her "spiritual body," as the apostle Paul calls it (more later on what our heavenly bodies look like), will go to heaven and her physical corpse will remain beneath the earth. With the coming of Christ, she will rejoin her body, and that body will be perfect, like the new earth itself—renewed and unblemished. When she was a little girl, Lotz told me one afternoon in her North Carolina office, she imagined that in heaven she would have a better singing voice and a smaller nose.

In 2001, Lotz wrote a slender volume called *Heaven: My Father's House*, in which she describes, in detail, the attributes of heaven. She compares arriving in heaven to the safety and comfort of coming home, finally, to a place where your father knows just what you want—just as she welcomes her own children home in special ways that show how much she loves them. "When my daughter Morrow comes home," she writes, "I know that she loves homemade chocolate pound cake, fresh flowers in her room, and time to look through home decorating catalogues." To imagine heaven is to imagine your own home, perfected—without leaks in the roof, faulty pipes, cracked paint, or threadbare rugs. To imagine living always in God's love is to imagine the love of the most perfect parent, who, like Lotz's own father, always kept the porch light on when waiting for one of his kids to come up the long, winding drive to their mountaintop home in North Carolina.

Lotz knows that I'm Jewish, and over lunch one afternoon, in an expensive restaurant in midtown Manhattan, she interrupted her discourse on heaven to witness to me about Jesus. "Lisa, God wants you," she said, her voice breaking and her eyes on mine. "You are precious to Him and you have a choice." I don't believe that my ultimate destiny has anything to do with Jesus, but Lotz's certainty made me squeamish. I looked down at my notebook and kept scribbling, unable to meet her gaze. I know she's wrong, I thought. But what if she's right?

THE BOOK OF REVELATION

Apocalyptic visions appeal especially to people seeking justice, who believe that their way of life and religious practice are under threat by the dominant mainstream culture. In a world divided into black and white, haves and have-nots, good and evil, heaven is both a reward and a consolation for righteous behavior (that is, for staying faithful under great pressure), and an exclusive club that separates "us" from "them." Whether here or in the hereafter, says L. Michael White, who is a professor of classics and Christian origins at the University of Texas at Austin, heaven helps those on the margins "form a distinctive identity as an ethnicity or a group." With vivid images of heaven—and especially the Judgment before heaven—"you define the in group not

only as Jews or Christians or Muslims, but us-good-Jews or Christians
or Muslims."

We are accustomed in modern-day America to relegate apocalyptic
beliefs to the white fundamentalists who over the past three decades
have comprised the religious right. But in the nineteenth century, some
of the country's most fervent end-time believers were American slaves.
Encouraged by the Bible's promises of justice—soon—and moved by its
images of radical upheaval, slaves sang songs that celebrated the world's
imminent collapse.

> And you'll see de stars a-fallin'
> And de world will be on fire
> And you'll hear de saints a-singin'
> And de Lord will say to de sheep
> For to go to Him right hand;
> But de goats must go to de left.

Slaves' songs and folk stories promised not only destruction, but
a heaven that rewarded them for their pain and humiliation in life.
Emily Burke, a white Northerner, observed this on a tour of Georgia
plantations. In her antebellum book *Pleasure and Pain: Reminiscences
of Georgia in the 1840s*, she wrote that slaves believe that "in the life to
come there will be white people and black people; but then the white
people will be slaves and *they* shall have the dominion over them."
The first, as the Gospels promise, shall be last.

Although America has proved to be fertile ground for apocalyptic
sects of all types, the fundamentals of apocalyptic belief—a margin-
alized group looking to God to set things right soon, certain that the
next world will be better and different in every way from this one—
have remained the same from the beginning.

Isaiah promised that a Messiah would come from the House of David to right every wrong, and Daniel and Enoch promised heaven to those Jews who remained, as they put it, "righteous." The book of Revelation is an heir to these texts—and the most argued-over book of the Bible. Martin Luther, the father of the Protestant Reformation, considered it "neither apostolic nor prophetic," and though he included it in his version of the Bible, he did so halfheartedly. (The first edition of Luther's New Testament, dated 1522, included Revelation in its table of contents—but Luther inserted a great space between it and the rest of the canon, as if to say, "I'm putting this here, but you can take it or leave it.") Nevertheless, Revelation is crucial to our story for three reasons. First, it contains all the images of heaven that have filtered into the popular imagination. Revelation has the angels singing "Holy, Holy, Holy." It has God on His sparkling throne, the white-clad saints singing praises, the gold streets, the verdant garden, the pearl gates, the flowing rivers, the jeweled walls. It has what is perhaps the best-known description of heaven anywhere: "God himself will be with them; he will wipe every tear from their eyes. Death will be no more; mourning and crying and pain will be no more, for the first things have passed away." Though the images in Revelation and its structure—a visionary travels to heaven and reports back on what he sees—are derived from generations of Jewish apocalyptic visions that came before, Revelation is the perhaps the most important text in the story of heaven because it is the one we all know—or think we know.

Second, Revelation offers a window into the mind of someone who truly believes that Christian capitulation to mainstream culture is about to bring God's punishment upon the world. Its tone is one of outrage over the pressure the Roman Empire put on all its inhabitants to participate in every aspect of public life and revulsion at those who called themselves Christians and yet embraced Rome's customs. Its prediction of cosmic consequences for those Christians who participate in Roman culture is as dire and unequivocal as Anne Lotz's words to me: there will be a fiery cataclysm—and soon. Repent or you will fail to merit God's beneficent judgment.

Finally, the book of Revelation seems to be written in code. The narrative thread is simple enough. A visionary introduces himself to his audience and tells them his story. He has seen wonders. Guided by an angel he has seen heaven, the throne of God, and God Himself, surrounded by angels. He has seen the saints and the righteous, standing before the throne and singing praises. He has also seen the future fate of mankind. God and his angels waged a great war upon the earth, unleashing terrible steeds and warriors to fight the devil, who appeared and reappeared in multiple forms: a dragon, a serpent, a multihorned beast. Everything turned upside down, "the sun became black as sackcloth, the full moon became like blood, and the stars of the sky fell to the earth as the fig tree drops its winter fruit when shaken by a gale." The author witnesses a final battle, the return of Jesus, and God's mysterious fusion of heaven and earth—or, as the text more exactly describes it, the descent of heaven to earth. "For the first heaven and the first earth had passed away, and the sea was no more. And I saw the holy city, the new Jerusalem, coming down out of heaven from God."

But it is the horror-movie images—Jesus returns wearing a robe dipped in blood; out of his mouth comes "a sharp sword with which to strike down nations"—and the symbolic numerology that have made the book of Revelation the ur-text for generations of end-time believers hoping to find within it the key to predicting the end of the world. The number seven appears again and again. Upon ascending to heaven, John sees seven blazing lamps and an unforgettable Lamb—with seven horns and seven eyes—"looking as if it had been slain, standing in the center of the throne." He sees seven angels blowing seven trumpets, and later seven angels carrying seven bowls (each bowl contains a plague). The devil in chapter twelve, a huge red dragon, has seven heads. Upon each head sits a crown.

Other numbers signal multitudes or greatness—"myriads and myriads and thousands of thousands" of angels are seen in Revelation, chapter 5. Two hundred million troops are sent forth to fight the battle against the devil, and witnesses make prophetic predictions for 1,260 days. Still other numbers signal threats and warnings: "Let anyone with understanding calculate the number of the beast, for it is the number of a person. Its number is six hundred and sixty-six." Hobbyists and prophets have long analyzed these numbers in various ways to come up with the exact dates of certain biblical events—the Creation story, the Flood, and the birth and crucifixion of Jesus—as well as of the end of the world. In the nineteenth century, William Miller, an American minister from Low Hampton, New York, used Revelation and an ancient Jewish calendar to predict that the world would end on October 22, 1844. (He had previously predicted that the world would end on March 21, 1843.) Today, Harold Camping, an octogenarian evangelist based in Oakland, California, with a radio show that reaches millions of households each week, predicts based on his own calculations that the world will end on October 20, 2011.

I met Leonard Thompson on a rainy afternoon in St. Paul, Minnesota, in a classroom of Luther Seminary. A professor emeritus at Lawrence University, in Appleton, Wisconsin, he was there for a meeting of the American Academy of Religion, and I'd come there to see him. Thompson has spent his life studying the book of Revelation, and says that after fifty years he still finds its images as moving as any he has ever seen. "Anyone who reads it would be gripped, not just by the terror but by these images. It creates drama, the feeling that there's something more there than meets the eye. I don't know any book that holds its power."

Conventional wisdom holds that the author of Revelation—who introduces himself at the beginning of the book as "John"—was imprisoned by the Romans on the island of Patmos, near the coast of modern-day Turkey, for preaching the Gospel. Scholars have all kinds of theories about who John really was and what he was doing in Patmos. "What I'm about to say is heresy," says Thompson: John was not a prisoner but a half-mad, itinerant preacher. Such travelers were common in those days; they would journey from house to house, telling stories in exchange for a meal and a bed. (Scholars understand this "John" to be entirely different from the John who wrote the fourth Gospel.) Revelation was written around 95 CE, and many scholars believe that John was probably a refugee from Jerusalem, a person who had seen the destruction of the Holy City by Rome twenty-odd years before and who, traumatized, considered the destruction of the city, and especially of the Temple, a sign of the last days. Most Christians in the first century, remember, were also Jews or of Jewish descent. They would have regarded the Jewish Temple as a holy place and its destruction as an abomination. Revelation, descended from the Jewish apocalypses and written by someone who was fluent in Jewish Scripture, reminds readers that in the new world created by God, there will be no need for any Temple, for God himself will reside among the righteous.

But what behavior by the first-century Christians made John believe that they were about to bring hellfire upon themselves? The answer, says Thompson, is sacrificial meat. In the first century, every inhabitant of the Roman Empire—even those who lived in far-flung outposts of Asia Minor, and to whom John was addressing himself—was required on certain feast days to sacrifice animals on altars dedicated to the Roman emperor. This ritual made the emperor equivalent to a god. On these feast days residents of the empire would pull couches from their homes into the street, fly flags off their porches, watch parades and athletic events, and drink and carouse into the night. These feast days amounted to a public display of state religion—like the Fourth of July if the president were God.

For the Christians (and the Jews, for that matter), the meat sacrificed on the altars to the Roman emperors presented an ethical problem, one that resonates with any member of a minority religion in America today. Could the earliest Christians participate fully in Roman society without violating the First Commandment? Could they, should they, make the required sacrifices to the emperor god? If not, could they still eat the sacrificed meat, which was sold at the market? Could they dine with their neighbors who served the meat for supper? Like American Jews, who make bargains with themselves over the degrees to which they will celebrate Christmas without eroding their Judaism—a Christmas tree but no star on top, Santa but no Christmas tree, stockings but no Santa, and so on—first-century Christians wrestled with conflicting priorities: their desire to belong versus the mandates of their religion.

John said no. No sacrificing of animals to the Roman gods or emperors. No eating of that sacrificial meat. No celebrating with the Romans in the streets, even if it was just for fun. Rome was a betrayer, a whore, the devil. In John's mind, capitulation to Rome amounted to the erosion of Christian identity and Christian truth. "Those who worship the beast and its image [the emperor and statues of the emperor, mainstream scholars say] will also drink the wine of God's wrath, poured unmixed into the cup of his anger, and they will be tormented with fire and sulfur in the presence of holy angels." The author of Revelation, according to White, "is saying to Christians, don't you dare do anything to honor this emperor. Don't you realize you're worshipping the very beast that destroyed Jerusalem? If you worship this emperor, you're actually worshipping Satan's henchman. If you go along with this, you're going to be destroyed. The righteous, those who resist, are going to be able to go to the New Jerusalem."

Thompson echoes White. "John of Patmos is saying, 'Don't associate yourself with the Roman world.' The eating of meat is a very real issue: if you affiliate with the Roman world, if you go to civic meetings, you'll have to eat the meat that's sold in the market." What is the alternative? I ask. Was the author of Revelation recommending a full-fledged revolt against Rome? "What he seems to say is lie low," Thompson answers. "God's going to do it all. Confess to being a Christian when asked by the authorities, but ignore them as much as you can. At the proper time, God will fight and destroy the Roman world with his holy angels."

Substitute "secular culture" for Rome and you understand the view of people like Lotz, who see everyday American life as a challenge to Christian belief. I heard her preach one cold and rainy Sunday at a Baptist church in midtown Manhattan. The church was filled despite the weather. Church ladies in big hats sat crammed next to disheveled twentysomethings looking worse for wear after Saturday night. Lotz's sermon that day was about Christian humility. Yet the subtext was entirely about the threat of popular culture to Christian belief. She preached against shallow, fashionable "paganism," against "gods and goddesses and rock crystals." She preached against what she sees as sexual immorality: "We call murder 'the right to choose'; we call fornication 'safe sex'; we call an abomination 'homosexuality.' " Christians who have been seduced by such relativism must repent: "It is time for God's people who are called by God's name to humble themselves and pray." And then, in case anyone had lost track of the stakes, Lotz reminded the crowd that "eternity is just a breath away. Eternity," she repeated in her gravelly voice, "Is. Just. Right. There."

THE KINGDOM AMONG US

Two hundred years after the death of Christ, the world still had not ended, and the Church Fathers began to work out an idea of "immanence." The Kingdom of God was not a real event about to occur in history, they said. The Kingdom *is* the world, transformed through the birth of Jesus Christ, his death and resurrection. For support, they looked especially to the Gospel of John, the most mystical of the four evangelists. "I have come as light into the world, so that everyone who believes in me should not remain in the darkness," says Jesus in the twelfth chapter of John.

Origen, the second-century Egyptian theologian, denied a physical resurrection and wrote about the coming of the Kingdom as a metaphor for the pursuit of spiritual perfection. In a meditation on the Lord's Prayer, he discusses the meaning of the phrase "Thy kingdom come." If we reject sin and perfect ourselves according to God's will, Origen says, we will literally be in heaven. "The Kingdom of God does not come observably," Origen wrote, "nor shall men say 'Lo it is here,' or 'Lo it is there,' but the Kingdom of God is within us."

In about 270, a wealthy Egyptian named Anthony, twenty years old and an orphan, went to church and heard these words from the Gospel of Matthew: "If you wish to be perfect, go, sell your possessions, and give the money to the poor, and you will have treasure in heaven." Inspired, Anthony went home and sold all his things. The next week, back at church, he heard these words: "Do not worry about tomorrow."

So Anthony found shelter in a burial ground near his property in a rural village named Coma, and for fifteen years fasted and prayed.

When he felt ready, Saint Anthony went out into the desert. He was, in some literal sense, going to the place where heaven touched the earth. In the desert, the sky, the stars, the planets, and the moon seem near at hand. The American Jesuit priest Mark Gruber spent a year in the Egyptian desert, following Anthony's footsteps, and in his diary, published as a book titled *Journey Back to Eden: My Life and Times Among the Desert Fathers*, he writes about the awesome omnipresence of heaven. "All my senses are overwhelmed," he writes, addressing God. "In my aloneness, in the purest solitude, you are achingly present."

In the desert, Anthony found a cave "full of creeping things," according to his fourth-century biographer Athanasius, and moved into it. There he wrestled with Satan, fasting and praying in an effort to purify his soul and grow more like God. He drank from a spring inside the cave, and he ate the bread that friends would occasionally drop down through a hole in the roof. In 1 Thessalonians, the apostle Paul had urged Christians to "pray without ceasing," and Anthony took this command literally—his prayer became as regular as the beating of his heart. Word of Anthony's extraordinary holiness reached the cities and towns of the empire, and pilgrims flocked to see him. He turned them all away. Finally, after twenty years, a determined crowd descended upon the cave and pushed a rock away from the opening. What they saw amazed them. Anthony "had the same habit of body as before, and was neither fat, like a man without exercise, nor lean from fasting and striving with demons," writes Athanasius. "His soul was free from blemish, for it was neither contracted as if by grief, nor relaxed by pleasure nor possessed by laughter or dejection." Anthony, in other words, had become perfect, like an angel in heaven.

Anthony was the first Christian hermit, and he inspired generations of imitators—known as the Desert Fathers. Throughout the deserts of the Middle East, solitary Christians performed feats of self-abnegation. One of the most famous, Simeon Stylites, climbed atop a pillar in the desert east of Antioch—and lived there for thirty-six years. Also in the third century, Anthony's compatriot Pachomius—who according to the fifth-century Christian historian Salminius Hermias Sozomen "was frequently admitted to intercourse with the holy angels"—established a desert community, a group of men, living together and praying, resting and working at designated times. This was the earliest monastery.

The Desert Fathers, these earliest hermits and monks, left behind little books of aphorisms and sayings. These men (and a few women) would have answered the "when do you go to heaven?" question in two ways. Soon, they would have said. They withdrew from the world because they believed strongly (like their predecessors the Essenes) that the end was coming. "The present age is a storm," said Amma Theodora, who lived in the Egyptian desert in the fourth century. At the same time, they acted as if they already *were* in heaven—without seeming to want to go there. The Desert Fathers said you can't strive—in any competitive, ambitious, or obvious way—to be like an angel. All you can do is pray to God constantly and endeavor in all humility to carry out His will (a concept almost Buddhist in its emphasis on selflessness). In one of the more famous stories from the Desert Fathers, one monk asks another what to do when he sees a brother falling asleep in church. Shouldn't he wake him up and command that he pay attention? No, says the older, wiser monk: "For my part, when I see a brother who is dozing, I put his head on my knees and let him rest." Heaven, in this view, is both something that will really happen, in history, and also a flickering light that one occasionally sees on earth, when people love each other the way God loves them, with perfect kindness, generosity, love, and compassion.

Perhaps you remember Franny Glass, the heroine of J. D. Salinger's *Franny and Zooey*, who, inspired by the account of a medieval pilgrim, endeavors to achieve enlightenment by saying the Jesus Prayer during a college-football weekend. Over and over, young Franny—a beautiful and brainy college sophomore—chants "Jesus have mercy on me, a sinner" to herself until, finally, she faints upon returning to her table in a restaurant after a trip to the bathroom. She has a nervous breakdown and moves home to her mother's Upper East Side apartment, where she sleeps and cries and snuggles with the family cat until her brother Zooey reminds her that having an ambition for sacred peace is the surest way not to attain it. "Jesus knew—*knew*—that we're carrying the Kingdom of Heaven around with us, inside, where we're all too goddam stupid and sentimental and unimaginative to look," says Zooey. "The Jesus prayer has one aim and one aim only. To endow the person who says it with Christ-Consciousness. *Not* to set up some little cozy, holier-than-thou trysting place with some sticky, adorable divine personage who'll take you in his arms and relieve you of all your duties." Zooey tells Franny that she had better go out in the world and use her God-given talent for acting, and that she had better do it as though she were performing for Jesus every night. The pursuit of beauty, plus humility and God's grace—that is heaven.

Monastic communities, then, were early examples of Christians living together in a conscious effort to replicate heaven on earth, and throughout the centuries such radical lifestyle experiments have been tried again and again. The Reformers continued to debate the question "What happens when you die?" Luther described the phase between death and paradise as sleep, which passes in an instant, but

Calvin insisted on something more like a conscious but disembodied being-with-Christ. When you died, Calvin said, you went to heaven. Other Protestant reformers agreed. "We believe that the faithful, after bodily death, go directly to Christ," reads the Second Helvetic Confession, written in 1562 for the Swiss Reformed churches, and adopted by Protestant churches throughout Europe. Yet the unreachable, majestic heaven given to Christians by the Reformers inspired them more than ever to create heaven on earth.

This country, argues Mark Holloway in his 1966 book *Heavens on Earth: Utopian Communities in America, 1680–1880*, was essentially founded by Christians seeking a home on earth for their own versions of heaven. These were communal societies, inspired by the accounts of the first-century church in Acts 4: "Now the whole group of those who believed were of one heart and soul, and no one claimed private ownership of any possessions, but everything they owned was held in common." The Puritans, of course, were hierarchical and autocratic. They lived according to stringent rules handed down by authoritarian ministers: the church was the center of their family and social life; the punishments of hell and the rewards of heaven were always supposed to be foremost in their minds. "The true END of Life" is "the SERVICE OF THE GLORIOUS GOD," preached the Boston minister Cotton Mather in 1726. Still, they saw their communities as exemplary models of Christian life, devotion, and piety. When John Winthrop, the first governor of Massachusetts, said, "We shall be as a City upon a Hill," he was talking about the new Jerusalem.

But other, more individualistic conceptions of heaven on earth flourished here as well. The Amish, an Anabaptist sect descended from the Mennonites that found fertile ground in America in the early eighteenth century, rejected modernity in an effort to live as much as possible like a community of saints in the world. The Ephrata Cloister, established in the Pennsylvania wilderness in 1732 and named after what the biblical Hebrews called Bethlehem, produced illuminated manuscripts and mystical poetry. Its members would wake for two hours in the middle of each night to await the arrival of Christ. The Shakers, who established communities in the late eighteenth and early nineteenth centuries throughout New England and in Kentucky, lived in austere houses, separated by gender. Sex was not allowed; children were thus procured only by adoption. They were called Shakers because when they became possessed with the Holy Spirit they trembled violently. They made wooden furniture so exquisite that the American Trappist Thomas Merton wrote in 1966 that "the peculiar grace of a Shaker chair is due to the fact that it was made by someone capable of believing that an angel might come and sit on it." The Oneida community in upstate New York experimented with "complex marriages," in which the community's men and women had sex on a regular, rotating basis. That way, no attachments would form, the theory went, and the members could devote themselves fully to Christ.

In 1840, the American transcendentalist Ralph Waldo Emerson complained in a letter to his friend Thomas Carlyle that the country was overrun with idealists, each chasing a utopian dream. "Not a reading man but has a draft of a new community in his waistcoat pocket," he wrote. "One man renounces the use of animal food; and another coin; and another of domestic hired service." By the early twentieth century, theologians such as Walter Rauschenbusch and, later, Reinhold Niebuhr were developing what came to be called "a social gospel." Niebuhr later repudiated that gospel. Discouraged by what they saw as the rampant and destructive individualism of American society and opposed to the working and living conditions of the poor in the newly industrialized American cities, these liberal Protestant reformers upheld what the *Encylopedia of Politics and Religion* calls "a faith in progress." They believed in "the perfectibility of human society and [had a] confidence that humans could bring about the Kingdom of God." Heaven, they said, happens when people treat each other on earth as Jesus instructed—as brothers and sisters. This faith in the Kingdom—now—translated in practical terms to armies of progressively minded believers founding urban organizations that fed the poor, housed the homeless, and tended to the sick. The civil rights movement of the 1960s was, conceptually, an heir to the social gospel.

Experiments in heavenly living continue to appeal to some. In 2003, a fiftyish teacher named Bren Dubay took a group of students on a field trip to Habitat for Humanity in Americus, Georgia. They had packed up the cars for their trip home when a Habitat worker urged Bren to take a detour, to Koinonia Farm—a place Dubay had never heard of, just seven miles away. "I had no interest in coming to see this farm," says Dubay. "I had things I had to do but, being a polite Texan, I said yes."

Koinonia Farm was founded in 1952 by a Baptist seminary graduate named Clarence Jordan who—inspired by the accounts of the earliest Christians in the book of Acts—hoped to build a Christian community around the principles of love, prayer, brotherhood, and service to others. The ideals of heaven were never far from Jordan's mind; he called Koinonia—named for the Greek word meaning "communion" or "community"—"a demonstration plot for the Kingdom of God." (A demonstration plot is where farmers experiment with new seeds or planting techniques—and then invite their neighbors to come see what they've done.)

It wasn't much—just a few little houses off a scrappy dirt road. There were a bigger house for meetings, some farm animals, and chicken coops. But from the beginning, Jordan's emphasis—especially on brotherhood and service—led the members of Koinonia to overthrow convention especially in matters of race. Jordan invited his neighbors—the grandsons and -daughters of the slaves and sharecroppers who had plowed that land for generations—to work with him at Koinonia and to share the fruits of that work. Jordan was "radical, into the social gospel," his friend Con Browne once said. So avant-garde were his ideas that at a family meeting, Jordan's parents raised the question of whether he should be committed to a mental institution. Only the opposition of a like-minded sister derailed that plan.

Koinonia became a target for the Ku Klux Klan. Members would awaken at night—many times, according to Browne—to find their houses on fire. Sometimes they would be shot at. "For one summer, no night passed without shots being fired nearby," Browne recalled. Georgia Solomon was a child who lived near Koinonia during that time and remembers that Jordan gave her pecans when she stopped by to visit. When she grew up, and had three babies and not enough to eat, the people at Koinonia built her a house. "I made it through my trials and tribulations," she said, "and now I'm striving for eternal life."

Membership shrank and grew through the years, from a peak of sixty people in the mid-1950s, down to just two families during the violent 1960s. In 1976, it spun off its house-building ministry into a little nonprofit called Habitat for Humanity. Jordan died in 1979. Today twenty-five people live at Koinonia, all in different stages of becoming members. Dubay is one of them. After that first visit—which was just forty-five minutes but "had such an impact on me"—she ordered Jordan's books and cassette tapes online. Later that year she went back again and this time stayed a week. In 2004, she left Texas for good and moved to the farm to become its director. Her intention is to bring Koinonia farm, after decades of internal and external strife, back to its first principles: love, community, and service to others. "Heaven and Jesus isn't about going to church for an hour or so on

Sunday and then way off in the future you go to heaven," she says. "It's about doing what Jesus asks us to do right now. We have a lot of visitors year-round, and we hear time and time again, 'This is a little piece of heaven for me.' Or 'This is the Garden of Eden for me.'

"It's beautiful here," she adds. "Some people would like for it to be more mountainous. Some people would like for it to have more water. But there's something sacred about the place." Under her leadership, the folks at Koinonia are fixing up the farm buildings little by little. They make their money, as they always have, from the fruit of their eleven hundred pecan trees—which they harvest and sell by the bag, salted and roasted, and make into pies and fruitcakes. It's not much, Dubay says, but it's everything they need.

FOUR

GREEN, GREEN PASTURES

It's eighty degrees in Silicon Valley and the sun feels hot enough to melt the pavement. Inside the headquarters of the Muslim Community Association, though, it's cool, and I am sitting in a nondescript, fluorescent-lit, windowless room with three men of devastating professional credentials. Laid out on a table, buffet style, is a lavish spread: chicken kebabs, vegetable soup, hummus, pastries, sweetened fruit salad. There's a pot of tea and a pot of coffee, signs designating their contents dangling like necklaces. The three men are listening respectfully as I explain my project, though they all have better things to do. Razi Mohiuddin, nearly fifty and born in India, is president and CEO of a small software company called Ironspeed. Hisham Abdallah, in his fifties, emigrated to the United States from Egypt twenty-five years ago; now he's a clinical researcher at the European pharmaceutical giant Roche. Mohammed Nadeem teaches e-business courses at National University in San Jose. What draws them all here is that they don't want to be lumped together with terrorists and fundamentalists. They want to set the record straight.

Except for the occasional uplifting or quirky feature piece (a profile of a halal butcher, for example), most Americans don't know enough about Islam to temper their news-watching with the awareness that, as a tiny fraction of the world's billion Muslims sets fire to the globe, the rest are going to work, coming home, and praying to God—leading lives not so unlike those of Christians and Jews. By various counts, there are anywhere from two to eight million Muslims in America, most of them immigrants or children of immigrants from South Asia, Africa, Central Europe, and the Middle East—but also a substantial number of converts, especially among African Americans and a growing number of Hispanics. Before 9/11, religion was something most American Muslims didn't talk much about, in part because they knew it could create tension between them and their employers, teachers, and neighbors. But since the word *Islamic* has become linked to the word *terrorist*, a generation of moderate American Muslims is "coming out of the closet," says Razi. "There's a new reason for my existence, if you will, which is to bridge the gap of understanding that exists today. Our kids are going through an identity crisis. Are they American? Are they Muslim? We have a job as parents to make sure that their identity is not one of conflict or contradiction."

Thanks to a handwritten note—part motivational screed, part polemic—that Mohammed Atta left in his rental car on the morning he flew American Airlines flight 11 into the Twin Towers, our cultural misunderstandings now include our ideas of heaven. Convinced he would die a martyr's death, Atta wrote about the rewards he would find in another world, dwelling especially on the companionship of ravishing women. After reminding himself to "strike [the enemy] above the neck," he writes: "Know that the gardens of paradise are waiting for you in all their beauty, and the women of paradise are waiting, calling out, 'Come hither, friend of God.' They have dressed in their most beautiful clothing." Naturally, the American media latched on to Islamic conceptions of heaven, focusing on the aspect of sexual companionship. The subtext of this discussion was usually: "Can you believe this stuff?"

The Qur'an does in fact promise to the male residents of heaven the company of beautiful maidens. The critical verses describe couches for lounging, young attendants bringing around cups of spring water and goblets of wine that don't intoxicate—and the company of houris, "eyes large and dark, like pearls in their shells, as a reward for past deeds." Few scholars believe that *houri* means anything but "virgin" or "maiden," in spite of a much-talked-about 2002 *New York Times* article by Alexander Stille, which quoted a Qur'anic scholar's theory that *houri* might mean, simply, "white," or even "white raisin." Some later Islamic texts, accounts of the sayings of Muhammad called "hadiths," promise a specific number of virgins to Muslim martyrs—sometimes seventy, sometimes seventy-two—but that is not a widely held belief. The seventy-two virgins were popularized in the Islamic world in the 1980s during the Iran-Iraq war; in recruiting tracts distributed to young Iranian men who would almost certainly meet their death in war, the highly sexualized virgins played a starring role. Moderates interpret the idea of heavenly virgins as a metaphor for sublime satisfactions unimaginable on earth—the way moderate Christians view gates of pearl and streets of gold. Such nuances were largely lost in the prurient editorializing. Even *Playboy* weighed in—"When will sex be delivered from the clutches of religion?"

Consequently, all three of my lunch companions are defensive as well as welcoming—understandably, since their sacred beliefs have been poked, prodded, and laughed at in public. They're thoughtful members of an affluent religious and social community—a sort of Islamic mega-church. There's a proper mosque down the road, but here—in buildings that formerly belonged to Hewlett-Packard—the Muslim Community Association offers a wider range of services. There are rooms for worship (separate blue-carpeted chambers for men and women), and a large hall that can seat six hundred people for dinner, as it does every year at the end of the holy month of Ramadan. As in Christian mega-churches, the community cares for its members' social, physical, and practical needs as well as addressing its spiritual concerns. It offers Qur'an study and Arabic classes, kickboxing and aerobics, free legal advice and a weekend health clinic. Four hundred children attend its elementary school, wearing uniforms exactly like those you'd see in a Catholic school—plaid pinafores over jeans and sneakers—except that in the older grades, many girls wear traditional *hijab*, or head covering.

Every Friday afternoon, three thousand men and women park in MCA's vast lot, pausing from their jobs as programmers, engineers, marketing executives, chemists, and CEOs to pray, in unison, the same prayers Muslims say all over the world: "There is no God but God." Afflicted by this economic downturn—house values have dropped by a quarter, and layoffs have hit the high-tech sector as everywhere else—MCA still represents one of America's most successful communities. MCA members earn, on average, about $100,000. They work at Yahoo! and Sun Microsystems, both just across the 101 freeway, and at dozens of other high-tech firms. They are first- and second-generation immigrants from at least forty different countries, including India, Pakistan, Bangladesh, Egypt, Saudi Arabia, Vietnam, Laos, Cameroon, and Bosnia. Many in the older generation have spent their lives working at hard, unglamorous jobs—they've run hotels and convenience stores—in order to send their sons and daughters to Harvard and Columbia. This community is the new American dream.

When I met him, Razi Mohiuddin was in his second term as the community's president. This is not a spiritual post but a leadership role, and Razi has a California-style grace and ease of manner that makes clear why he's a popular choice. He's tall, with an athletic build, a wide smile, thick black hair and mustache. Razi has lived in the United States since he was eighteen, when he arrived in Chicago from India, alone, to start a degree program in computer science at the University of Illinois. Middle age and prosperity have made Razi more, not less, religious. He prays five times a day, as mandated by the Qur'an, and sends his children to the school at the mosque. Because the Qur'an forbids borrowing money with interest, Razi paid cash for his house. It's easier to be a practicing Muslim, he notes, when you're the boss: he schedules meetings around his prayers. But his two identities—practicing Muslim and successful software entrepreneur—are sometimes at odds. He agonizes when he has to throw a party for his sales force: the guys want beer and wine, but how can he justify such an expenditure when the Qur'an explicitly forbids the consumption of alcohol? He'll do it, but not easily. "I take you out to dinner, I buy you a glass of beer—there's something in my heart that has a hard time being happy about that."

For Razi the idea of heaven is always present—he cannot imagine his life without it. Like most faithful Muslims, he believes that God watches him constantly, and that his every action—from washing his hands in the morning to his final prayer at night—is being recorded in a big book by God's angels. According to the Qur'an, each one of a person's deeds is inscribed in heaven. "All they do is noted in the Book of Deeds. Every matter, small and great, is on record." The hadiths flesh out the details, describing a book, or a scroll, that gets fastened to your neck as your soul departs your body. On Judgment Day, Allah hands down verdicts: the righteous receive their book in the right hand, the doomed in the left.

The Qur'an also speaks of an enormous balance, or *mizan*, that weighs a person's good and bad deeds on the Day of Judgment. "They whose scales are weighed down—these shall prevail," the Qur'an says. "They whose scales are light—these have lost their souls, and in hell shall abide for ever." Those who come up short, in other words, will go to hell, a place of "scorching wind and boiling water / And a shade of black smoke, / Neither cold nor kindly." Those with a positive balance sheet will go to heaven.

Razi accepts the Qur'an's descriptions of the pleasures offered in Paradise as true—the maidens, as well as the upholstered couches, the ripe fruit, the flowing fountains and running rivers—but beside the point. The promise of heaven is a constant reminder: Do right by God by saying your prayers; do right by others by treating them with kindness and respect. According to Islamic tradition, Razi says, you can ask a person whom you've wronged for forgiveness at the end of time—and if forgiveness is granted, both the seeker and the granter go to heaven. The problem, Razi explains, half laughing, is that you have to find that person among all the souls who ever lived on earth, all huddled together waiting to be judged. It's simply easier to treat people well—and to ask their forgiveness when you don't—here and now. To emphasize his point he e-mails me, several days after our meeting, a local news story. A Los Angeles cab driver, a Muslim, found $350,000 worth of diamonds in the backseat of his taxi. He tracked down the passenger who had left them there and gave them back. "God is up there," the driver said by way of explanation. "He always watches."

ORIGINS

The desert where Islam was born is 750 miles from Saint Anthony's Egyptian desert and much more brutal. This is the Hijaz, a stretch of sand on the western coast of Saudi Arabia, relentlessly hot—in summer, daytime temperatures average 113 degrees—and by outsiders' standards uninhabitable. At the end of the nineteenth century, the British explorer Charles Montagu Doughty described this desert thus: "The sun, entering as a tyrant up on the waste landscape, darts upon us a torment of fiery beams, not to be remitted till the far-off evening." In the sixth century, this desert was inhabited by Bedouins, nomads who raised animals, traded skins and meat for produce and weapons, and broke camp constantly in search of new land for grazing. We know—because the Qur'an condemns the practice—that parents sometimes buried newborn girls alive in the sand so as not to have one more person to feed. In this desert, the only thing that mattered was survival, and the trick to survival was holding on to your land as long as you needed it and relying on your kin to help you protect it. There was no comfort here.

Ancient Bedouin religious practices were loose; religion was not an organizing or motivating force in Bedouins' lives. The idea of paradise as a reward for the righteous would have been laughable. Still, they did have religious traditions, inherited from their ancestors and practiced out of habit and pragmatism. They believed in a variety of gods, but not in any organized pantheon or hierarchy. There were gods of places and gods of weather—of rain and sky. There were gods in trees, in animals, and especially in stones. Like the ancient Hebrews, the Arabs sacrificed animals to those gods on stone altars and ate the flesh. Like Hindus, they sometimes gave an animal—usually a beloved camel—to the gods, consecrating it without killing it. In a society where camels literally gave people life—meat, milk, transportation, and shelter—it is not surprising that they would regard certain of those animals as holy, separating a chosen one from the herd and bestowing upon it a retirement of rest and ease.

The Bedouins did have a sense, if an inchoate one, that the dead "went" somewhere—or why else would they lame a camel, tie it up next to the grave of someone recently deceased (a heroic warrior, for example), and leave it there to starve? The camel must have been thought to provide transport to another world. But, in general, life was life and death was death: inevitable, lamentable, often unfair. Illiterate and unschooled, these nomads left behind heart-stopping oral poetry, passed down from generation to generation. When the subject was death, as it often was, the poet sang of revenge, of grief, of finality— but never of heaven. "I will cry for you with unsheathed swords," one poet sang, "with sharpened saber, and with lances. These are the arms of he who must avenge your blood." Another poet—one can almost imagine it was someone's grieving mother—keens, "Why does death persecute us so? Every day, she takes a noble man from our tribe. / She has a preference for the best of our sons. She chooses the most virtuous and most illustrious." These are not the words of a person who hoped for anything better at the end of her own life.

Situated in the middle of this desert, at the bottom of intersecting gullies, was the only slightly more habitable town of Mecca—a place of "suffocating heat, deadly winds, clouds of flies," according to Maqdisi, a tenth-century Arab geographer. In the summer, the town grew so hot that people spoke of the city as "burning." Clusters of mud-and-straw houses huddled together in the center of town atop what was essentially a drainage ditch. During the season of torrential rains and flash floods, these houses were regularly washed away. Mecca's biggest—some contemporary scholars say its only—attraction was the Ka'aba: a plain, ancient building in the center of a large square near a big market. A large black stone, probably a meteorite, was set into its eastern corner. For the nomads—as well as for the residents of Mecca and for the merchants and traders who passed through on routes north to Palestine, south to Yemen, or west across the Red Sea to Egypt— that stone had long had religious importance. According to tradition, Abraham, the Old Testament Hebrew patriarch, had stopped in Mecca to visit Ishmael, the son he had with his mistress Hagar, and he himself built the first rough temple around the stone. Mecca became known both as a holy place and as neutral territory—a place where feuding tribes could lay down their arms, a place where everyone made sacrifices to the god of the Ka'aba. That god was called "Allah": not the only god the people who worshipped there knew, but certainly the most important. For centuries, pilgrims had traveled there to pray, in a ritual they called the hajj.

Western scholars are in the midst of radically revising their views on the social and historical circumstances that gave rise to Islam. For the last half of the twentieth century, this field was ruled by the great British scholar W. Montgomery Watt. His position, considered the mainstream view, is that by the end of the sixth century, Mecca had become a prosperous and populous city. It was home to merchants, but also to moneylenders and creditors, and to the organizers of great caravans that took leather, spices, gold, silver, grain, arms, fabrics, oil, and perfume north to the Mediterranean and south to Yemen. A wealthy elite emerged: people with the means to send their infants into the cooler areas of the desert to be nursed by Bedouin maids and their families there for summer vacations. This sudden wealth, Watt and his camp say, changed the nature of Arab society. The tribal code of blood gave way to a mean-spirited individualism. The tribal loyalty demanded by conditions of scarcity was replaced by a less equitable system, in which the haves gave and received preferential treatment while the have-nots suffered unfairly.

You could argue, then, that the social conditions that produced the Qur'an and Islam were not so unlike those that produced the book of Daniel in the first place—and the Gospels two hundred years later: dramatic social upheaval created a yearning for a new kind of justice. But this argument is, in practice, nearly impossible to make. Watt's view cannot be confirmed because almost no one in sixth-century Arabia could write. There were no record keepers, no diarists. Historians can only rely on the Qur'an and other early Islamic texts, which—as we surmise by analogy from contemporary New Testament scholarship—probably contain some truth but are biased in favor of the new dominant religion. Even to study such biases is problematic, though, because of the Qur'an's special place in Islam. The traditionally faithful believe that the Qur'an exists exactly as it is, in heaven, and was given to Muhammad by God in smallish doses, over a period of two decades. There were no editors, no translators, no interpreters, no process of canonization, no mutations over time—and to discuss the possibility of alternate truths amounts not just to scholarly disagreement but to blasphemy. Unlike the Bible, the Qur'an is not a narrative. It contains few psychologically complex characters, and even fewer stories. It is a divine revelation that came straight from God. It is in itself holy. Generations of scholars have observed that the proper analogy for the Qur'an is not the New Testament but Jesus himself. But a new group of scholars, especially in the West, has begun to take a colder look at the Qur'an, trying to separate fact from myth, fully aware that their efforts may create extreme discomfort, if not outright anger, among even the more moderate of the faithful.

This newer view has less narrative drama than Watt's, although the conclusion—that a destructive stratification of society took place—is the same. Mecca rose to prominence, the new thinking goes, because the dominant tribe there, the Quraysh, had the political savvy to figure out how to capitalize on the Ka'aba, to draw merchants and traders away from the trade routes and into the town. Mecca, in this view, was not a bustling center in itself, but a must-make detour off the highway.

Islamic scholars have long argued that Mecca was home to "pagans and idolaters," and that Islam brought the revelation of monotheism to the ungodly, but that perspective is also under renovation. As we have seen, Allah, the god of the Ka'aba, was growing ever more powerful. And although Meccans had not taken up Judaism or Christianity, representatives of those religions must have been passing through the city with increasing regularity to visit the holy site. According to the traditions, hundreds of idols and icons surrounded the Ka'aba, including paintings or statues of Mary and Jesus. Some traditions say that images of Mary and Jesus had been painted directly onto the interior walls of the Ka'aba. (According to certain of these accounts, Muhammad placed his hands over these images when he conquered Mecca and ordered his followers to erase everything else; Jesus, in the Islamic tradition, is a holy man.) Certainly, there were well-established Christian and Jewish communities within a camel ride of any intrepid Meccan. Christians had settled to the south of the Hijaz, in Yemen and Ethiopia, as well as to the north in Syria. In the Hijaz itself, there were powerful Jewish tribes just two hundred miles away, in Medina. One of them, the Qaynuqa, could field an army of seven hundred men, about half in full armor. And in Mecca itself, an indigenous monotheism was flowering. The *hunafa* were local intellectuals who found Judaism and Christianity insufficiently monotheistic—they were especially suspicious of the Christian Trinity—and who secretly practiced what they called "*din Ibrahim*," the faith of Abraham. This, they said, was a pure faith; the others were corruptions. In other words, Mecca may have been pagan in a narrow sense, but the seeds of monotheism were circling in the hot desert air, looking for a fertile place to land.

Into this world Muhammad was born, near Mecca, probably around 570. He was the son of a merchant family, and an orphan. He was raised by an uncle, and was, according to the hadith, a sensitive, introspective, and spiritual child. When he was about twelve, he went with his uncle to Syria, where he spent several days in deep conversation with the ascetic Christian monk Bahira—apparently a formative experience. As a young man, he spent considerable time away from Mecca, in the mountainous regions of the desert, praying, thinking, reflecting in solitude. At twenty-five, he married a wealthy widow named Kadijah, who was, as legend has it, old enough to be his mother—but since they had at least six children together, she was probably only in her thirties at the time of the marriage. Watt surmises that Kadijah, an independent businesswoman with her own means, could afford to marry Muhammad for his spiritual, rather than his material, prospects.

Accounts of divine revelation appear again and again in the Western world's holy literature. Moses, John of Patmos, Anthony—all were humble men, alone with their thoughts, with no expectation of being singled out. In the year 610, Muhammad had gone on retreat to a desert cave and was sleeping when a voice came to him. It is one of the most gripping moments in sacred scripture, telegraphed in the Qur'an and elaborated upon in the hadiths. Gabriel, God's angel, woke him up, gave him a document, and commanded him, "Recite!" Startled and confused, Muhammad asked, "What shall I recite?" (His words have also been interpreted as "But I cannot read," emphasizing the miraculous origins of the Qur'an.) In response, the angel physically squeezed him so hard that Muhammad felt he could not breathe, then commanded again: "Recite!" And again Muhammad asked, "What shall I recite?" And the angel said, "Recite, in the name of your Lord! / He Who created! / He created man from a blood clot. / Recite! Your Lord is most bountiful. / He taught with the pen. / He taught man what he knew not." Muhammad recited.

After this encounter, he ran home to his wife in terror, told her his story, and said he was afraid he was losing his mind. But Kadijah believed in him. She was his first convert, and slowly the other members of his household, and then many residents of Mecca, began to accept Muhammad's message—that Allah was the One God, the beneficent,

the merciful. In 622, Muhammad moved with his small band of followers to Medina, where conditions were slightly more hospitable. (For
one thing, there were more Jews in Medina, and they were more ac-
cepting of Muhammad's monotheism.) From 610 until Muhammad's
death in 632, God spoke regularly to His Prophet, and Muhammad
preached God's word to the Arabs of the Hijaz—in some of the most
eloquent, lyrical poetry in history. Those verses comprise the Qur'an,
and to the more than one billion believing Muslims in the world, they
are truly the word of God.

The Qur'an establishes Islam's "five pillars"—the five things every
Muslim must do to please God. The first pillar is similar to the Bible's
First Commandment. The Islamic version says, "There is no God but
God" and establishes Muhammad as God's true messenger. The other
pillars are: praying five times daily, giving alms to the poor, fasting
during the month of Ramadan and, if possible, making the hajj, or pilgrimage to Mecca. If the Qur'an can be said to have a single message,
it is that to please God, people must behave rightly and responsibly,
both as individuals and within their communities. As Karen Armstrong puts it in her best-selling synthesis *Islam*, "The old religion . . .
was simply not working. . . . The way forward lay in a single god and a
unified [community] which was governed by justice and equity."

PARADISE

Islam's merciful, beneficent God would reward his faithful with paradise, and the paradise described in the Qur'an must have been a vision
beyond any desert dweller's imagining. For one thing, it was temperate.
Verses refer to "green, green pastures," "cool pavilions," and "fountains
of gushing water." The food and drink in Mecca and its surrounding
desert must have frequently spoiled, for the Qur'an promises a heaven

where everything tastes fresh: "In it are rivers of water, not brackish, and rivers of milk, unchanging in taste, and rivers of wine, delicious to them who drink it, and rivers of honey, pure and limpid. Therein they shall enjoy all kinds of fruits, and forgiveness from their Lord." Residents of paradise will feast on a variety of fruits—pomegranates are specifically mentioned—and "flesh of fowl, whatever they desire." In paradise, men who had spent their lives toiling beneath a burning sun and fighting over small patches of sand will wear gold bracelets and green robes and recline upon upholstered couches, while servants pass around goblets of cool water and wine. In paradise, people greet each other with the word *salaam*, or "peace."

(Tradition says that Arabic is the language of heaven. Ten years ago, the Penn State scholar of Islam Jonathan Brockopp was browsing in a Cairo bookstore when an old sheikh with a long white beard befriended Brockopp's wife, Paula, who was waiting in the street. "Do you speak Arabic?" the sheikh asked. "A little," said Paula, whose father is a Lutheran minister. "That's good," said the sheikh, half teasing, "because when you get to heaven, you'll want to know what's going on.")

But first, the world must end. According to the Qur'an, a day will come when the stars will fall from the sky, the sun will stop shining, the ocean will boil (reminiscent of the language of Revelation). At that hour, the Qur'an says, paradise is near.

Allah himself resurrects the dead—first Abraham, Moses, and Muhammad, then everyone else—and all crowd before Him, standing shoulder to shoulder, awaiting His Judgment. How long is a matter of some debate. Some traditions say a thousand years, others say fifty thousand. Naked, terrified, sweltering, the resurrected and soon-to-be-judged have ample time to contemplate their sins. According to some traditions, they can become so anxious that they wind up standing up to their necks in sweat.

Finally, the Judgment is handed down. The Qur'an talks about a scale in which God literally weighs a person's good deeds against the bad and metes out sentences accordingly. Those bound for heaven receive their "book," the record of their life, in their right hand; those bound for hell receive it in their left. Then everyone turns to walk over a bridge as thin as a razor's edge—an adaptation perhaps from Zoroastrianism. The just ones breeze over it, guided by God, into the gardens beyond. The damned fall off into the stinking pit, where "garments of fire have been tailored / and over their heads is poured scalding water / melting therewith their innards and their skins."

At first Muhammad's message of heaven and hell fell on disbelieving ears. In verse after verse, the Qur'an depicts the locals doubting, ridiculing, and dismissing ideas of resurrection and the afterlife. "We shall not be resurrected," they say. The God of the Qur'an deals squarely with these doubters: "O mankind, if you are in doubt about the resurrection / We created you from dust, then from a sperm, then from a blood clot, then from a morsel, formed and unformed, to make it plain to you. . . . And you will see the earth lifeless / But when We send down the rain upon it, it vibrates, and doubles its yield / And comes out in plants, of every kind, a joy to behold. This is so because God is the Truth. / It is He Who revives the dead, and has power over all things." When this message did finally take hold—and with it a message of personal responsibility to God and community, writes Jane I. Smith, a scholar of Islam at Harvard University, in her book with Yvonne Haddad, *The Islamic Understanding of Death and Resurrection*, it helped the Arabs value their role, both individually and collectively, in history as it unfolded according to God's will. This evolution, from indifferent plural gods to a beneficent and caring Allah, gave the nascent Muslims an appetite for making improvements to their own world and the world around them.

Conventional wisdom teaches that the Islamic heaven is more material and vivid than Jewish or Christian versions. Its pleasures are more physical and its attributes—the landscape, the garb, the society, the pastimes—less disputed. "For Muslims, the body isn't a problem," the Islamic religion professor Kevin Reinhart told me when I visited him one slushy winter day at Dartmouth. "The things of this world are anticipations of the world to come: good conversation, the pleasure of sitting around with your friends, good textiles, and wine that doesn't give you a headache." Like all simplifications, Reinhart's assertion is both true and not true.

Professor Khaled Abou El Fadl has made it his mission—or you could say vocation—to disabuse the world of the notion that a billion Muslims can believe any one thing. Born in Kuwait, he was educated at Yale University and now works as a professor of law at UCLA. Emulating the great Muslim intellectuals of the Middle Ages—the eighth-century orthodox theologian Hasan al-Basri and the twelfth-century classical philosopher Ibn Rushd—Abou El Fadl has created what amounts to an influential progressive Muslim think tank. He surrounds himself with graduate students—Muslim and non-Muslim, American and foreign born—who are committed to the classical Islamic values of rigorous argument, diversity of opinion, and prolific publication. He has written scholarly books on such topics as the Islamic view of beauty and the role of women in the Muslim world. His wife, Grace Song, is a convert from Catholicism and the daughter of Taiwanese immigrants, who has devoted herself to being her husband's scribe and amanuensis. Abou El Fadl himself is the great man at the center of it all: he is chronically ill, he doesn't drive, he rarely knows the time—or even the day of the week. His commitment is solely to discovering truths and disseminating them.

What Americans don't understand about Muslim heaven, he told me one afternoon in a long phone conversation, is a lot. Too many people retain the image "of someone killing themselves so they can go have a sexual orgy with a bunch of virgins." Far more troublesome to him, though, is the literalism with which so many Muslims, both worldwide and in America, read the Qur'an. Over the last hundred years, and especially in the last thirty, what Abou El Fadl calls an Islamic "Puritanism" has become the dominant form of the religion worldwide—thanks in large part to well-funded evangelism efforts coming out of Saudi Arabia. This Wahhabism, as it's called, emphasizes strict (Abou El Fadl would say uncritical) adherence to religious and lifestyle rules and a belief in the Qur'an as divine that makes it impervious to historical or cultural contextualizing, let alone critique. Such a worldview has created a global community of Muslims opposed to enlightenment or progress, he says. "The laity knows Islam through media and public schools—and of course from the Wahhabis themselves. A lot of the representatives of Islam in the United States know practically nothing about the Islamic tradition. . . . We're living through the dark ages of Islam." The Muslims at what Abou El Fadl calls this "ABC level" connect their rewards in paradise with their actions on earth the way children behave at their grandmother's house because they'll get ice cream later—because, as the scholar puts it, "I want to be able to eat sweet apples in the hereafter."

By not engaging with Islam's history and its classical tradition, Abou El Fadl believes, American Muslims bear some responsibility for global terrorism, and he said as much in an editorial in the *Los Angeles Times* the week after 9/11. The inward-looking nature of Puritan Islam as it is taught throughout the world "produced a culture that eschews self-critical and introspective insight and embraces projection of blame and a fantasy-like level of confidence and arrogance." For such statements as this, he has received death threats from fellow Muslims. Former friends no longer speak to him. He is not welcome at many of the country's Islamic centers. (While researching Abou El Fadl one day at the NYU library, I discovered a Facebook group campaigning to invite him to that campus. The Facebook petition called him "the most important and influential Islamic thinker in the modern age," and lauded his insistence on open dialogue among Muslims. One Facebook member was critical. "*La hawla wala qowata illa billah*," he wrote. "There is no power or strength but Allah.")

When Abou El Fadl speaks of his own belief in heaven, he speaks not of fruits, or sex, or "green, green pastures." He points to other passages in the Qur'an and the hadiths—those more inscrutable and demanding. Abou El Fadl likes to contemplate what the Qur'an means when it says that God's "throne was upon the waters," a phrase that shows God at once in and beyond the world. He likes the verse in which God asserts that He is so close to the people under his dominion that He is "nearer to him than his jugular vein." He especially likes the famous verse 24:35: "God is the light of the heavens and the earth / His light is like a niche in which is a lantern / The lantern in a glass / The glass like a shimmering star . . . Light upon light! / God guides to his light whomever He wills / And strikes parables for mankind / God has knowledge of all things." Light upon light is a beautiful phrase. "We don't know what that means, but it's worth exploring," he tells me.

"Like all humans, I've struggled with paradise. The thing I am at peace with: there is accountability. There is a Creator. There is consciousness and memory. I've long abandoned any attachments to any notions that are bounded by space and time because they philosophically don't make any sense. Even the texts of the Qur'an don't make any sense. God keeps sending us these prophets who keep telling us there is more than material existence—and then after material existence there's more material existence. That makes no sense. I do believe in a world of souls. It must have tranquility, repose. But all that physical stuff stopped being convincing a long time ago." I cannot forget the image from Paul Barrett's 2007 book *American Islam.* Barrett, a journalist, visits Abou El Fadl at his small suburban home lined with floor-to-ceiling books in every room but the bathrooms. At prayer time, Abou El Fadl, his teenaged son from a previous marriage, and Song stand shoulder to shoulder on small prayer rugs—no separation of the sexes here—and rise and fall, prostrating themselves: the woman, the man, the child, all born in different corners of the earth, chanting quietly in Arabic, the language of heaven.

VIRGINS: A FEMINIST PERSPECTIVE

Muslims—especially those born abroad like Abou El Fadl—cannot understand the American obsession with the houris, with the idea that Muslim heaven must be something like a sex party for boys. They laugh at our insistent questioning, and with laughter imply that this whole orgy-in-heaven thing is our hang-up, not theirs. Our questioning is not wholly unenlightened, however. "Muslims who dismiss the significance of the belief in the houris of paradise are perhaps oblivious to feminist concerns," the Vanderbilt University historian of Islam Leor Halevi wrote to me in an e-mail. "They haven't really thought about the lack of a comparable sexual reward for heterosexual women. Alternatively, they interpret the material delights of paradise in a highly metaphorical way, as symbolic indications of spiritual rewards applicable equally to women and men." As proof that questions about houris, sex, and women's entitlement in

the next world are not merely twenty-first-century American obses-
sions, Halevi—with help from Dartmouth's Reinhart—points me to
a fourteenth-century Persian poem, which clearly reflects the same
concerns:

> A preacher was saying one day
> "The houris of fair Paradise
> Provide a delightful surprise:
> Each man will have twenty, they say."
>
> An old woman rose from her place,
> "And are there men houris as well?"
> "If you show your face there," he said,
> "You'll not escape when they give chase!"

Both sexes, in other words, can avail themselves of heaven's sensual
delights.

Laleh Bakhtiar doesn't think heaven has anything at all to do
with sex, as we humans know it. She was born in Iran and raised in
Los Angeles and Washington, D.C., by a single mother who was a
Christian. She went to Catholic school and was confirmed in the
Roman Catholic Church. She did not begin her investigation of her
own Muslim roots until she went to Iran to visit her father's family
when she was in her twenties. Now seventy, she has been a Muslim
for more than four decades. She raised three children in the Muslim
tradition and for five years taught the Islam section of a comparative religion class at the University of Chicago's Luther Seminary.
She has written numerous books on Islam and specializes in the
Sufi tradition.

Bakhtiar was toiling in relative obscurity when she gained the attention of the world with her 2007 translation of the Holy Qur'an. In it, she used gender-neutral language, and she softened some of the Qur'an's most egregiously sexist passages—notably the controversial Sura 4:34, which most scholars agree condones wife beating. (That sura, according to a more conventional translation, says, "Banish them to their couches, and beat them." Bakhtiar makes it: "Abandon them in their sleeping places; and go away from them.") In her introduction to *The Sublime Quran*, Bakhtiar writes that she hopes this translation will be of use especially to American women who have been stuck, as it were, using conservative or traditional translations their whole lives—and so have internalized language that in effect made women into second-class citizens. The Qur'an has been translated and interpreted exclusively by men for fourteen hundred years. "If women start interpreting the Qur'an you get a completely different picture."

Imams around the world renounced Bakhtiar's translation, banning it from use in their mosques and congregations; in response, other scholars drafted statements supporting it. The Islamic Society of North America—whose president is also a woman and a convert—upheld Bakhtiar's translation and noted that Bakhtiar was neither the first nor the only Muslim commentator to conclude that wife beating was not Qur'anic.

In any case, even Bakhtiar—whose feminist orientation might lead one to suspect a cynical view of the delights of heaven—laughed at the suggestion that those pleasures might be exclusively for oversexed men. Paradise, she says, "is for everybody. The Qur'an does describe a place with pillows and cushions and grapes, but it is certainly not a sex place." But what, I ask, about the houris, so universally accepted as maidens or virgins, comely and available? Bakhtiar circumvents the fundamental sexism of Muslim heaven by asserting that the word *houri* has no gender. In her translation she calls them "lovely eyed ones, black-eyed." "People may have thought of them as female, but they're not necessarily," she says. Besides, she adds—as almost all Muslims do when confronted with questions about physical descriptions of paradise—"your body is not there. We're in a completely different form. We're not going to feel the same things there. I'm not thinking about what kind of orgy we're going to have in paradise." She prefers a metaphorical interpretation, in other words: her vision of her life after death is not so different from Abou El Fadl's. "It would be my hope," she says, "that I would be part of the light that would shine in the world or the universe." Sex is not part of her picture.

Back in that dim room in the Silicon Valley office park, Hisham Abdallah, the Roche pharmacist, is telling me how Muslims get to heaven. There may be nearly as many interpretations of the Islamic heaven as there are Muslims, but on the general outline of the journey from grave to garden, nearly everyone agrees, because, as he says, "the Qur'an is like a road map or a manual." Abdallah is self-taught as a Qur'anic scholar. He is a lively narrator; at one point he compares the razor-thin bridge that spans the fires of hell to the narrow bridge of Khazad-dûm from which the wizard Gandalf falls in *The Lord of the Rings*. He walks a fine line himself, acknowledging that while the Qur'an depicts heaven in fantastical, cinematic detail—no ineffable or inscrutable abstractions here—only simple people interpret the accounts of those bounties literally. To fixate on the pleasures of heaven—the girls, the couches, the wine—is "not wrong," he says, "but it's incomplete. . . . Those things are promised, and they are there, but all we know are their names. Companions or fruits or meats or houses—anything you hear about in the hereafter, you will never be able to comprehend the nature of it. It's important to appreciate heaven for the deeper meanings, not just a depiction of the pleasures." Thus the wine in heaven won't have anything to do with what we know as "wine." It will be wonderful, Abdallah explains; it will enhance your sense of fun, but you won't necessarily have to uncork the bottle or drink it from a glass. "Heaven is a new form of existence which we cannot explain, but it's a fuller existence than what we have now. The hereafter is the real existence," he says.

Abdallah says he thinks about heaven every day, especially in the morning, when he first awakes. In Islamic literature, death is often compared to sleep, and the afterlife to a reawakening. Every morning, he says, he recommits himself to God. "If you do everything with the intention of pleasing God, then this whole thing we call life is about the hereafter."

Razi Mohiuddin, the software entrepreneur, is much more straightforward. He believes wholeheartedly in paradise as a place of unrestricted abundance, and every heavenly symbol in the Qur'an—the couches, the girls, the wine, the fruit, the flowers, the fountains—reminds him of that plenitude. The economic woes of recent years make him even more eager for the rewards of paradise: Today, "I may have sufficient money to go to a restaurant, but I'll hold back because there's a rainy day I may have to plan for. . . . Over there, one will hopefully not have these restrictions at all. There won't be any rainy days to worry about."

Every time Razi sees heads of state posing for a photo op, with a bowl of fruit on the table between them, he thinks about heaven. Every time he goes to a wedding and sees a bridal bouquet, he thinks about heaven. These small tokens, he says, are constant reminders of the world beyond, of the rewards bestowed upon those who pray, give to charity, and do right by others. "Gardens and rivers and fruits are the best of what we have in this world. Does that mean heaven will be exactly like [it says in the Qur'an]? I don't think so. It's a place where I can do effectively whatever my desire wants, without getting penalized for it in any shape or form. Everything is acceptable. There is an abundance of everything. There is harmony with people, with things, with nature. There is the ultimate peace of mind." And then he puts the keys to the community center back in his pocket, glances at his cell phone, and heads back to the office.

FIVE

RESURRECTION

On the morning of September 11, 2001, Dr. Betsee Parker was waiting for the locksmith. An affluent woman, the kind some would call "horsey" (she organizes exhibition polo matches and trains "hunters and jumpers" for competitions), Parker was finally about to accomplish a chore that long needed seeing to: getting the lock fixed on the door to the guest apartment in her city house, on the Upper East Side of Manhattan. Trained as an Episcopal priest, Parker was on hiatus from professional religious life, happy enough to live the life of a society matron. But when the elevator door opened and she saw the expression on the face of the locksmith, she knew something was wrong. She put on her clerical collar and took the subway downtown. She wanted to see how she could help.

In the ensuing weeks, Parker found a peculiar vocation. It became her job to unzip the body bags as they arrived at the makeshift morgue at the medical examiner's office on East Thirtieth Street, lay her hands on whatever was inside, and say a blessing. She wanted to relieve the emergency workers—who would stand, silently, around the mortuary table as she prayed—of one horrible duty, she told me, and to provide some kind of comfort to them if she could. The quality of the remains, to use the lingo of the medical examiner's office, was awful; more often than not the contents of the body bags resembled sludge more than anything human. She went through many trousers pockets, discovering what she could there—a wallet, a photograph of a child. At night, she would look up from her work at the endless row of ambulances, lights flashing, all waiting in a queue to unload more of their gruesome freight. Her colleague, another Episcopal priest, Dr. Charles Flood, told me that one victim was found in over one hundred pieces; another was scattered over two acres of lower Manhattan. Thanks to the miracle of modern DNA tests, scientists could identify the fragments' common source. But they could not put the body back together again.

Often, and especially on those nights, Parker would think about heaven. "I would say to myself, 'Someday I will see these people. They will be perfected. I will be there and I will meet these people who died in this terrible catastrophe, and the rescue workers too. In the resurrection, the body and soul are one.' "

What happens to your body in heaven? Do you believe, as Parker does, in physical resurrection? That is, do you believe that at the end of time, your body, some version of the one you have now, will be reunited with your spirit and that you will live forever, looking recognizably as you did in life? This question, more than any other question about heaven, puts the modern believer to the test. The twentieth-century poet Czesław Miłosz, who was Roman Catholic, wrote about the difficulty of understanding resurrection this way: "So I believe in an absurdity, that Jesus rose from the dead? Just answer without any of those evasions and artful tricks employed by theologians: Yes or no? I answer, Yes, and by that response I nullify death's omnipotence." It was one thing to believe that God raised corpses from the dirt and made them whole back in the days when people also believed that bloodsucking leeches cured disease and that the sun and the planets circled the earth. It's much more difficult to continue to believe in this kind of magical process when we know what we know about matter and its inevitable decomposition, and (despite Frankenstein fantasies) about the blunt impossibility of making what is dead live again. For me— personally speaking—resurrection is the biggest obstacle to belief in heaven. As much as I'd like to imagine that in heaven I'll get my pre-baby body back—its pleasures, its stamina—I just don't. We get old, we wear out, we die. Time never moves backward.

My mother recently had surgery for colon cancer, and as she was recuperating one afternoon, she surveyed her body, seventy-two years old, which had seemed to deflate all at once with the operation. "You look down and you don't even recognize this body as yours," she said to me, with a look of surprise in her dark eyes. My mother was a staggeringly beautiful young woman, the kind of woman who attracted looks from men in restaurants well into her fifties. I would like to believe that when she dies she will look in heaven the way she did in the 1970s, dark and glowing, with glossy skin and silver hair—as unself-conscious in her beauty as any woman ever was. But I just don't.

I am in good company. Only 26 percent of Americans believe they'll have bodies in heaven, according to a 1997 *Time*/CNN poll. And if you need further evidence of a weakening belief in resurrection, nearly 30 percent of people in a 2003 Harris poll said they believed in reincarnation—and of self-professed Christians, that number was 21 percent. Reincarnation and resurrection have, traditionally, been mutually exclusive. Orthodox Christians would find the thought that anyone among their number was privately harboring hopes of coming back to earth in someone else's body to be not just illogical but heretical.

The best proof that resurrection belief is fading, however, is the rise in cremations in America. Thirty years ago, almost no one chose to be cremated; now almost half of Americans say they'd prefer cremation to burial. Cremation has long been discouraged, especially in conservative religions—among the Mormons and the Eastern Orthodox, for example—as a practice that desecrates the body, which is God's creation, and denies resurrection. Southern Baptists—who are more likely than the general population to believe that the end of time is at hand—are the least likely of those surveyed to cremate their dead. Still, 16 percent of them prefer the option, according to a 2007 report from the Cremation Association of North America.

Even the Roman Catholic Church has recently changed its position on cremation. According to Catholic theology, the human body is the sacred vessel for the Holy Spirit. For that reason, funeral masses were always said over an intact human body—in a coffin, in a church. But since 1997, acknowledging that economics or family circumstance sometimes make cremation the best or only option, Rome has waived that restriction for Catholics in North America. It's better to say a funeral mass over a body, the Vatican maintains, but ashes will do in a pinch.

Jewish, Christian, and Muslim traditions continue to teach that our bodies are sacred; our bodies, together with our souls, make us who we are—individual parts of God's miraculous creation. "The Lord God formed man from the dust of the ground, and breathed into his nostrils the breath of life," reads the passage in Genesis, "and the man became a living being." But we are likelier these days to try to claim our physical perfection in this life, and we shrug over the question of our bodies in the next. "It used to be," remembers the Michigan undertaker and poet Thomas Lynch in a phone call, "that couples four times their married weight would go to the parish fair and dance around each other twice a year. They knew that in heaven they'd be back to their best bodies. Now, with our emphasis on vitamins, face-lifts, and health care, we want our best bodies now." My friend Stephen Prothero, author of *Religious Literacy* and a professor of religion at Boston University, frames the problem more succinctly: "It seems fantastic and irrational that we're going to have a body in heaven."

In truth, the religious have always been divided on the question of flesh versus spirit in heaven, which scholars frame as an either/or: "resurrection or immortality of the soul." The orthodox (Jews, Christians, Muslims) insist on resurrection. Without this supernatural event, they maintain, belief in heaven is child's play, too namby-pamby to have any theological seriousness. Progressives circle the resurrection question cautiously. In their desire for "rational religion," they say that the word *resurrection* means something other than what it seems to mean, that it is a metaphor for a new kind of life. They fall back on that great theological cop-out: "We cannot know what God has in store for us." For my part, I prefer the rigors of the orthodox view—even if I can't embrace them. The surest, straightest path to belief in heaven is a belief in God's literal ability to revivify flesh. Everything wonderful about the traditional view of heaven follows from there: the reunions, the banquets, the gardens, the cities, the learning, and the loving. How could any of this exist without a physical self to enjoy it?

Resurrection is central to the Christian story—it's what Easter celebrates, after all. Among classical and modern Muslims, too, belief in resurrection has always been, as the scholars say, "normative." The Qur'an leaves no doubt about it:

> Yes indeed!
> I swear by the Day of Resurrection!
> Yes indeed!
> I swear by the soul that remonstrates!
> Does man imagine We shall not reassemble his bones?
> Indeed! We can reshape his very fingers!

That bodies and souls will be reunited at the end of time "as whole, cognizant, and responsible persons" goes without saying, write Jane Smith and Yvonne Haddad in *The Islamic Understanding of Death and Resurrection.* "Upon these realities there is nothing in the Qur'an or any other Islamic writings—scholastic or devotional—to cast the slightest shadow of doubt." The sensual pleasures of the Islamic heaven—food, drink, silken clothing, gold jewelry, good conversation, comfortable furniture, and even sex (or something like sex)—are utterly dependent on bodies with which to enjoy them.

Orthodox Jews also insist on a physical resurrection at the end of time. The twelfth-century Jewish philosopher Moses Maimonides made resurrection the last of his thirteen nonnegotiable tenets of the faith—and Orthodox Jews affirm that belief in a daily prayer called the *Ani Ma'amin*. "I believe with perfect faith," says the prayer, "that the dead will come back to life when God wills it to happen." It is for this reason that a minority of religious Jews follow a traditional prohibition against cremation and organ donation: they want their bodies buried whole to make less work for God at the end of time. Some Orthodox Jews even bury their loved ones with soil from Israel, a symbol of the time when they will rise up, finally in the Promised Land, to greet their Messiah.

I began this chapter with the Reverend Parker and the 9/11 victims because ever since the idea of resurrection took hold in monotheistic religion, it has strained credulity. The question of how God can reassemble the bits and pieces of a human who has been eaten by lions or—as so many continue to be in war—vaporized or pulverized is critical: if you can believe in a God who can reassemble a man who died in two hundred pieces and send him to heaven, whole and beautiful, then you believe in a God who can do anything. For such believers, heaven is no problem. But for the rest of us, who may think of

God more abstractly, skepticism about resurrection can feel like the barrier to entry.

RESURRECTION IN JUDAISM

Scholars fight bitterly over the thirty-seventh chapter of the book of Ezekiel, written during the Babylonian exile. In it, the prophet Ezekiel is looking over a great valley, which is filled with dry bones. The Lord puts his hand on the prophet and asks him a question: "Mortal, can these bones live?" You can almost see Ezekiel shrugging, when he says he doesn't know. Then God raises the bones and puts them back together, bone to bone, the way my nephew snaps together his Legos. Upon the bones, God puts sinew, then muscle and flesh. Ezekiel is impressed, but not fully, for the bones remain dead. Then God tells Ezekiel to command the winds to breathe life into the bones and he does so—and the bones "lived, and stood on their feet, a vast multitude."

God proceeds to tell Ezekiel what this miracle means: "These bones are the whole house of Israel. They say, 'Our bones are dried up, and our hope is lost; we are cut off completely.' . . . O my people; and I will bring you back to the land of Israel. . . . I will put my spirit within you, and you shall live, and I will place you on your own soil; then you shall know that I, the Lord, have spoken and will act, says the Lord."

What is God talking about here? What kind of promise is He making? Orthodox Jews insist that this verse is the ultimate biblical evidence for resurrection, a belief so central to Jewish tradition that it is impossible to be Jewish without it. Progressive Jews, on the other hand, say that God is talking about a figurative resurrection, not of individuals but of a people, a tribe, a nation. In exile, the Jews are despondent, cast out of their homes, and living far from their Temple and the burial caves of their ancestors. With these verses, God is promising that He will bring them back, alive, to their homeland where they can live in fulfillment of His covenant. He is not talking about magic, the progressives say, but about the collective future of a people.

Jon D. Levenson belongs to the first group. He is a natty Harvard professor who wears a tweed cap and a bow tie and has *New Yorker* cartoons plastered all over his office door at the Harvard Divinity School. Levenson is a man on a mission. With his book, *Resurrection and the Restoration of Israel*, Levenson, a traditionally observant Jew, wants to bring the belief in physical resurrection back to mainstream Judaism. Too many modern Jews, he argues, are content to abandon their tradition's insistence on resurrection.

I met him on the frigid morning in 2007 when fall turned into winter. He was late—and irritated—because his boiler was broken and no one had yet come to his house to fix it. I went to see him because I am literal minded, and I hoped he could explain to me how—faced with the realities of modern physics—he could believe in a God who could raise the people squashed in the Twin Towers, or incinerated, or blown to bits in war. Levenson's dark eyes twinkled. "Rationalist questions are not helpful," he said. "It's no use to ask, 'If I had a lab at MIT, how would I try to resurrect a body?' The belief in resurrection is more radical. It's a supernatural event. It's a special act of grace or of kindness on God's part."

His argument is simple enough. To the progressives—like my friend Alan Segal, who says that a Jewish belief in resurrection is nowhere in the Bible—Levenson counters, it is. The Hebrew Bible contains all kinds of hints and intimations of resurrection, Levenson argues: the Temple in Jerusalem is a place of restoration, purification, and renewal. The idea that a life-giving God would eventually restore humans to Eden is embedded in the rabbis' understanding of the Torah and the

Jewish tradition. During the Second Temple period, around the time of the heavenly promise given by Daniel, Jews began to understand that God would intervene and reverse the order of things—and that some would live again. "You always had a god who was on the side of life. Death, which is the most natural thing in the world, would be overcome."

Since before the birth of Christ, Levenson notes, observant Jews have been chanting their conviction that they would live again, in their bodies, in a new world after the end of time in their daily prayers. The most observant Jews say the Amidah—probably codified in the decades after the Romans sacked Jerusalem in 70 CE—three times a day, four times on the Sabbath. The prayer praises God, who is mighty and powerful—and "who causes the dead to come to life" (m'chayei ha-meitim).

But in 1824, members of a synagogue in Charleston, South Carolina, concerned about the inroads Christian missionaries were making into their community, convened a meeting and established the beginnings of Reform Judaism in America. As a result of that meeting, they revised Maimonides' thirteenth principle, replacing the belief in physical resurrection with "the immortality of the soul." More than six decades later, America's Reform rabbis met, this time in Pittsburgh, and introduced more amendments, including a disavowal of the kosher laws, which, they said, were "entirely foreign to our present mental and spiritual state." They also abandoned any residual Jewish belief in heaven, hell, and resurrection. The soul is immortal, they said, the spirit divine, but "we reject as ideas not rooted in Judaism the beliefs both in bodily resurrection and in Paradise and Hell as abodes for everlasting punishment and reward." Thereafter, the wording of the Amidah in the Reform prayer book was also changed. No longer did Reform Jews praise God who "causes the dead to come to life." Instead, they praised God "who gives life to all" (*m'chayei ha-kol*). The rabbis, in other words, obliterated resurrection from the contemporary Reform understanding of the afterlife. In 2007, the Reform movement revised its siddur, or prayer book, again—and this time the rabbis reinserted the ancient avowal of resurrection as an alternative reading to *m'chayei ha-kol*. This decision was debated for years, one of the editors once told me; Reform scholars and rabbis didn't want to be seen as retreating from their commitment to rationalism. Most Reform Jews still see resurrection as a metaphorical, not a literal, thing.

The debate over the place of resurrection in modern Judaism comes as news to my friend Rebecca. Raised in a Jewish enclave in Montreal, educated through high school in Jewish schools and summer camps, Rebecca is more at home in her religious tradition than most of my friends. Her parents belonged to an Orthodox synagogue when she was a little girl; in school, she studied the *Tanach*—all the canonical Hebrew Scripture—as well as commentaries on the *Tanach*. Nothing in her Jewish education had anything to do with resurrection, she tells me. No one ever brought it up. "It was not about looking forward to what God was going to give us. There was no discussion of heaven, of hell, of what was going to happen when I die. Ever. I don't think we even talked about the soul very much." As she remembers it, the flood came and obliterated everyone and everything on earth but Noah and the animals. Game over. "The emphasis," she tells me, "was always on what we could do in this world." When Jews like Rebecca have no sense of resurrection, it would seem that Levenson has a lot of work to do.

1 CORINTHIANS 15

The history of Christianity has been, in a sense, a two-thousand-year argument with those who disbelieve the literal truth of the resurrection.

And from the beginning, another idea has offered skeptics an alternative. "The immortality of the soul" has given them a way to believe in some kind of existence in heaven without having to buy into revivification. In the fourth century BCE, Plato taught that the body was a trap, a downward drag on the soul. The soul's goal in life was to tame the body's desires—hunger, thirst, lust—and so attain wisdom. Death set the soul free, and the wisest of souls would ascend to the highest places in heaven to dwell with God. Plato's rationalistic approach to questions of God and heaven has influenced every important Christian theologian since Clement of Alexandria, the second-century theologian and educator who said the Greek philosopher was "under the inspiration of God." Plato made it possible to talk about mystical abstractions in a reasonable way and so made theology a serious branch of philosophy. "Among the pupils of Socrates the one who shone with a glory so illustrious that he entirely eclipsed all the others, and not, indeed, unworthily, was Plato," wrote Augustine. "He was an Athenian by birth, of honorable standing among his countrymen, and he far surpassed his fellow students in his marvelous natural talent." In the Middle Ages, Thomas Aquinas and his fellow Christian intellectuals—as well as their counterparts in Judaism and Islam—looked to Plato and, increasingly, Aristotle as models for carving rational pathways through complex questions about God, the nature of evil, and the afterlife.

Around the time of the birth of Christ, most Jews believed in resurrection in the broadest sense. The Essenes, as we've noted, believed that the end of the world was imminent and that a resurrection—or, at least, a spiritual ascension—would accompany that end. They imagined that righteous, faithful Jews would return to the new world as angels—not flesh and blood, but celestial creatures who dwelled with God. The Pharisees, the precursors of the rabbis, believed something like what Jesus taught and what the Reverend Parker believes: humans, having been created by God, would be made perfect after death, in some combination of spirit and body.

Only the Sadducees, the elite and powerful priests who vied with the Pharisees over control of Jerusalem's Temple and its court, rejected resurrection entirely. In the Gospels they tease Jesus about it, asking him about a woman who had seven husbands in life: Which is her husband in heaven? (If you have a body in heaven, they're saying, then you must also have a spouse and—by implication—sex with that spouse.) Jesus becomes irate. "You know neither the Scriptures nor the power of God," he says. "For when they rise from the dead, they neither marry nor are given in marriage, but are like angels in heaven." Astonishing vistas are in store, he says—you would know this if only you read your Scripture.

The Jews of the first century expected a Messiah, as we've seen. According to the book of Chronicles, the Messiah would be a priest or a king, descended from the house of David. Jesus did not fit the prophecy. He was not from the house of David, and was not a king but a poor carpenter from Galilee. He was crucified, a humiliating, public, inglorious death. Yet when his followers rolled away the rock from the burial cave on the third day after his death, and the body of their teacher was not there, they drew only one conclusion: that he had been resurrected.

The task of the Gospel writers, then, was to convince their audiences that the resurrection had *really* happened—for it was one thing to believe in a future resurrection and quite another to believe that it had been a historical truth. It was not a metaphor: the man did not simply disappear. He was reborn. Physically. The disciples saw him. Each of the Gospels takes up this challenge differently. In Matthew, Mary and the other women see their risen Lord on the road to Galilee and they touch him: "They came to him, took hold of his feet, and worshiped him." In Mark, the resurrected Jesus appears three different times: first to Mary Magdalene, then to two of the disciples, and finally to the disbelieving eleven. In Luke, he appears to the disciples in disguise. He goes home with them. He blesses bread and hands it around (see—he has a body after all), and then he reveals himself. In John, of course, Jesus offers the ultimate proof. He places Thomas's hand upon the wound in his side. "Stop doubting," he commands, "and believe."

The apostle Paul mounted a particularly vigorous defense. Faced with audiences of pagan skeptics—for whom resurrection was not just incredible, but laughable—Paul launched arguments and explanations that were, at once, imaginative and derisive. In his letter to the people of Corinth, in modern-day Greece, he addresses his pagan skeptics as if they were just too dense to understand the importance of what had happened to Jesus. "Someone will ask, 'How are the dead raised? With what kind of body do they come?' Fool! What you sow does not come to life unless it dies."

The body is like a seed, Paul explains. Its properties in the afterlife will be the same as in life—yet totally different. "What is sown is perishable, what is raised is imperishable. It is sown in dishonor, it is raised in glory. It is sown in weakness, it is raised in power. It is sown a physical body, it is raised a spiritual body." This new body has superpowers. The miracle of resurrection conquers—it reverses—death and decay. " 'Death has been swallowed up in victory.' 'Where, O death, is your victory? Where, O death, is your sting?' " This passage is perhaps the most famous and most quoted explanation of physical resurrection in the Bible.

Paul then delivers the promise that has comforted so many in grief. "Listen, I will tell you a mystery! We will not all die, but we will all be

changed, in a moment, in the twinkling of an eye, at the last trumpet."
We will be there. We will be changed. It is a mystery. I think again
of my beautiful mother, and I hope.

Jim Caraley has seen too much grief. The central tragedy of his life
is now more than thirty years old, and still he cries when he remembers it. He agreed to see me one afternoon in his office near Columbia
University, to tell me about how 1 Corinthians 15—where he found
the passage he translates as "Oh, stupid man!"—saved his life.

In July 1979, Caraley's nineteen-year-old son, Christopher, a brash
and charismatic boy, vanished in a mountain-climbing accident.
Christopher had lied to his parents that summer; he said he was
meeting friends in Glacier National Park, a massive ice field in southwest Canada. Instead, he struck out alone, leaving his extra gear with
two French climbers he'd met at the ranger station. He was planning
to make a quick ascent of two peaks called Mount Athabasca and
Mount Andromeda, he told them, and would return for his things.
After a week, the Frenchmen alerted the park rangers and the rangers
launched a search. They also called his dad.

In New York City, the father picked up the phone. He was a young
professor of political science and had just returned from a trip to
Athens. "You're telling me my son is dead," he remembers saying to
the ranger. We're not sure, the ranger replied. Caraley flew to Alberta
and convinced the rangers to take him up in a search-and-rescue heli-
copter. "I could see the peaks and crevasses and the ice on the mountains, and it was very scary, and hard to believe that anyone could
scale them and come back alive," he says. The rangers never found a
body, and later that summer, Caraley wrote his son's obituary for the
newspaper.

On Thanksgiving, Caraley got behind the wheel of his car and fantasized about driving into a tree. In his anguish, he turned to the

Bible: it was something to distract him from himself. Raised in the United States by Greek-speaking parents and baptized in the Greek Orthodox Church, Caraley applied himself to reading the New Testament in its original Greek. But nothing in it gave him any comfort until he arrived at 1 Corinthians 15. Somehow, Paul's explanation allowed Caraley to believe that his beautiful Christopher had his body in heaven—and, better, that Caraley himself would see and recognize his boy at the time of his own death—"in the twinkling of an eye, at the last trumpet." Caraley responded especially to Paul's "senseless man" rebuke, as if the apostle could see him in all his grief and anger, in the rationalism that was driving him crazy. Paul taught Caraley that resurrection was a miracle: "I got myself to believe that," Caraley says. "If I hadn't, I would have gone mad. I would have hired a crew to chop down that mountain." Having absorbed Paul's lesson, Caraley calmed down. He remembers, a year after Christopher's death, being on a flight from Boston to New York so turbulent that the attendants were lying facedown in the aisle and his seatmate was saying her prayers. Caraley felt nothing but calm. He knew that if the plane went down, he would see his son again in heaven.

Levenson would prefer that I not dwell on rationalistic questions, but in fact such questions have obsessed theologians for millennia. How exactly would God do it, the earliest Christian thinkers wondered, and when He did, what would be the result? (According to a legend popular since the third century, the Christian martyr Perpetua has a vision of her dead younger brother Dinocrates, whose disfigurements in life vanished in the next world.) What would people look like in heaven? What body parts would they have? Which ones would they use? These questions particularly nagged at the Church Fathers, the men who in the first four hundred years of Christianity laid the philosophical and conceptual groundwork for what was to become the

Roman Catholic Church. The early Christian martyrs presented an especially thorny problem, for these men and women often died (like the victims of 9/11) in bits: burned, ripped in half, eaten and digested by animals. The Church Fathers agonized over minutiae: if an animal ate a martyr and then a human ate that animal, the second-century author Tertullian wondered, how would God find and reassemble the human particles? If one was eaten by a cannibal, how could God separate that individual's tissues from those of his host?

What age would we be in heaven? Would we have gender? Would we have hair, beards, fingernails? Would they grow? What about genitals, digestive tracts? In her brilliant and enduring 1995 book *The Resurrection of the Body in Western Christianity*, the Princeton scholar Caroline Walker Bynum sums up the thinking of Tertullian: "Mouths will no longer eat, nor will genitals copulate in heaven," she writes, for eating and procreating are part of an ultimately corrupt process of biological change. In heaven, "mouths will sing praises to God," and genitals "will survive for the sake of beauty. We will not chew in heaven, but we will have teeth, because we would look funny without them."

THE BISHOP OF HIPPO

But the person who most influenced our contemporary understanding of physical resurrection was Augustine, the fourth-century African Bishop of Hippo, who almost single-handedly laid down the theological foundations of the Western church. The son of a Christian mother and a pagan father, Augustine was a wild, unmoored youth who had a Christian conversion experience at the age of thirty-two. Thereafter he—prolifically—defined Christian thought in treatises, letters, and books, most of which (five million words, including at least five hundred sermons, all written down, by hand, by scribes) survive today. His masterwork, *The City of God*, which he finished in 426, just a few years before his own death, affirms not only the truth of the resurrection but defines, in satisfying detail, the nature of the resurrected body.

As a recent convert and a young priest, Augustine, like so many intellectuals, embraced resurrection halfheartedly. He had lived much of his misspent youth among Manicheans, a heretical Christian sect that had a Platonic dualism at its core: the body was a useless weight; at death, the spirit soared free of its chains. In 387, the year after his conversion, Augustine wrote that after death the soul yearns for "flight and escape from this body here below."

Over time, though, his ideas changed. As a working pastor, Augustine could no longer preach abstractions; he had to console his flock—real people, who really feared death. Perhaps even more important, he—like the writers of the Gospels and the apostle Paul—had to convince doubters of the truth of the resurrection, and in his world (North Africa in the waning days of the Roman Empire) half his neighbors were pagan and thought the idea laughable. In *The City of God*, Augustine "squared the circle," says the Boston University Augustine scholar Paula Fredriksen in her lovely, excited voice during one of the many phone conversations I had with her on this topic. Augustine blended the Jewish understanding of history, as something horizontal that ends and yet endures even after the end, with the Platonic understanding of the soul as something that goes up, vertically, toward God.

At death, Augustine wrote, the soul ascends, leaving the body to rot. The place to which that soul ascends has some correlation to the person's life on earth. Faithfulness to God, "righteousness," goodness, wisdom—all these things matter for your deeds on earth determine your hierarchical place in heaven. At the end of time there is a Judgment in which bodies and souls will be reunited. Resurrection is real.

In heaven, bodies are bodies, but they are also perfect. Augustine's argument to skeptics was like Levenson's argument to me. God will resurrect the faithful dead, even those whose death makes them unrecognizable, "devoured by beasts, some by fire, while some perish by shipwreck or by drowning in one shape or other, so that their bodies decay into liquid." God can do miracles. God is God.

Then, in one of the most delightful passages about the afterlife ever written—*The City of God*, book 22—Augustine wrestles with the rationalistic questions, one by one. People likely rise, he writes, with the bodies they had when they were thirty, the age Jesus was at his death. If they were too fat, they become thin. If too thin, they become heavier. If people were—at birth or by later accident—deformed or out of proportion, those deformities would be taken away, although martyrs would probably retain the marks of their martyrdom as heavenly badges of honor, just as Jesus showed the disciples his stigmata. Baptized babies rise in the form they would have taken as adults. (Unbaptized babies, he said in one of his most controversial teachings, would go to hell—though the torments they experienced there would be so mild, they would barely notice a thing.) Women rise as women, men as men. Indeed, in the resurrection women are more beautiful than they are in life, but their beauty elicits no lust. There will be no sex, no eating, no marriage (as Jesus said specifically). The soul will remember the body's former physical pleasures, but intellectually, without regret or temptation. Augustine describes this condition thus: "When the body is made incorruptible, all the members and inward parts which we now see assigned to their various necessary office will join together in praising God; for there will then be no necessity, but only full, certain, secure and everlasting felicity."

And what do these perfect bodies do in heaven? They see God, even with their eyes closed. They are swifter than birds, they go where the soul wants to go, they experience existence without sin. Time unfolds like one continual Sabbath. In heaven, writes Augustine, "we shall be still and see; we shall see and we shall love; we shall love and we shall praise; behold what will be in the end, and without end. And what other end have we except to reach that Kingdom which has no end?"

Wayne Ridgley, whom I met through a chaplain at a Veterans Administration hospital in Tampa, Florida, has never heard of Augustine, but he knows as surely as he knows anything that when he gets to heaven, he'll have his perfect body back. It's far from perfect now. He was a teenage marine in 1967 on night patrol in Vietnam when a mortar blew off most of his right leg and took a roast-beef-sized chunk out of his thigh. "I got shot, blown up, and stabbed," is how he puts it.

Ridgley is not just crippled, he is also addled—as he is the first to admit. He wears a high-tech prosthesis, but even so he is often in terrible pain. He cannot walk downhill without holding on to something—a railing, a cane—or stepping slowly sideways so as not to lose his balance. I arranged to meet him, some of his friends, and the VA chaplain for lunch at a Perkins Family Restaurant near the Museum of Science and Industry in Tampa. Ridgley showed up an hour late because the sound of the traffic got him disoriented and he lost his way. When he arrived he was too unraveled to talk for a while, or to order food.

Ridgley knows that God saved his life that day. "I believe in God all the way," he says, with a smile on his handsome face. "God is my guy." In heaven, he says, "our legs will be right, and since it's God I'm sure He can work a miracle and get our brains right, too."

"You get a new body," agrees his friend Ed Sitton. "Women don't have to get a breast augmentation because some jerk says that's how he wants them to look, and I won't be limping anymore." Sitton was on river patrol in Binh Dinh Province one August night in 1968, when he and the ten other guys in his boat were attacked. "I knew we were all going to die. Logic and common sense tell you that when you're so outnumbered, you're not going to live." As luck—or God—would have it, Sitton did not die. His ankle was shattered, though, and he has spent the rest of his life with a hobbled gait and worse—a paralyzing case of posttraumatic stress. (When I tried to catch up with him later, the chaplain told me he'd left Tampa, and friends thought he was living off the grid in the woods in South Carolina.)

Neither Sitton nor Ridgley has any interest in an Augustinian heaven where you praise God all day. Ridgley wants both legs working so he can play baseball—and win. Sitton points out, coyly, that perhaps winning won't be a priority in heaven, since there'll be no competition there. Ridgley waves that objection away. Once a marine, always a marine—and marines play to win. "Tough rocks," he says, and smiles.

SALVATION

In the early summer of 2008, Barack Obama—still politically wounded from revelations that his erstwhile pastor, the Reverend Jeremiah Wright, was an anti-authoritarian, left-leaning loudmouth— convened a meeting in Chicago with some of the nation's most important conservative Christian pastors and leaders. It was a diplomatic mission: Obama wanted to convince social conservatives that he was not too radical to be president. Billy Graham's son Franklin was there. So was Steven Strang, a conservative Christian publishing tycoon, and T. D. Jakes, the wildly popular African American preacher. According to news reports, Graham confronted the presumptive Democratic nominee with the question that separates conservative evangelicals from everyone else: "Do you believe Jesus Christ is *the* way to God or merely *a* way?" Obama, who knew that the gun was loaded, answered carefully. "Jesus is the only way for me. I'm not in a position to judge other people."

When my colleague Richard Wolffe and I interviewed Obama for a *Newsweek* profile later that summer, we circled back to the subject, and Obama was more expansive. He believes that there is more than one path to heaven, he told us, or else his beloved mother, who was agnostic, would not be there. "I've said this before, and I know this raises questions in the minds of some evangelicals. I do not believe that my

mother, who never formally embraced Christianity as far as I know . . .
I do not believe she went to hell." (He was quick to add, again, that
for himself Jesus Christ was the only way.) And although he believes
in his personal salvation through Jesus, he also believes that it's what
you do on earth that matters in heaven. "I am a big believer," he told
us, "in not just words but deeds and works." He understands that these
views make him unorthodox in the eyes of some. "My particular set of
beliefs," Obama noted, "may not be perfectly consistent with the beliefs of other Christians."

How do you get to heaven? The question of salvation is the biggest
wedge issue among the faithful. It's really a double-barreled question:
First, do you believe that the path you're on—Christian, Muslim, or
Jew—is the only path to heaven? Franklin Graham does—that's why
his query to Obama was so potentially explosive. And second, what
kind of Christian, Muslim, or Jew do you have to be to get there? Do
Orthodox Jews gain heaven, but not the intermarried? Do Sunni Muslims get to paradise, but not the Shia? Do you believe that "being a
good person" will get you into heaven? Or do you believe that you get to
heaven through God's mysterious will, no matter what you've done—
or not done—in life? Scholars call this the Works versus Grace debate,
and all three monotheisms have their versions of it. The answers affect
everything from relations between neighbors to the peace of the world;
they have motivated crusaders, martyrs, ascetics, and social reformers
through the ages; they are at the heart of the culture wars being waged
in America today. Claiming to know what God wants can have devastating consequences.

No answer seems fully satisfactory. If you're in the "deeds" camp,
then *which* deeds (and how many) will do the trick? Is it enough to do
unto others as you would have them do unto you, honor your parents,
and stay away from your neighbor's spouse? Or do you also have to go

to church, pray in specific ways, give money to charity and time to the poor? And if you satisfy all the aforementioned requirements, can you then jaywalk or drink four glasses of wine at dinner? Versions of such dilemmas triggered the Protestant Reformation, and they persist in American popular culture, often in the form of humor: A man stands before Saint Peter and tells him he never went to church, never gave to charity; he led, in sum, a completely self-centered life. Saint Peter is exasperated. "Have you never done *anything* good?" he asks.

"Well," says the man, "I once saw a group of Hell's Angels steal a purse from an old lady and as they were shoving her around, I intervened, took the meanest Hell's Angel by the collar, and told him he was despicable. And then I spat in his face."

"Wow," says Saint Peter. "When did that happen?"

"About two minutes ago," says the man.

The shortcoming of the "works" view, of course, is that it places eternal destiny entirely in an individual's hands, a perspective that anyone with a smidgen of experience of the world can tell you is laughable—and which, as Martin Luther discovered, leads to the most heinous abuses. By the end of the fifteenth century, the Roman Catholic laity was so consumed with assuring for themselves and their loved ones a speedy salvation that they were paying cash to their church authorities in exchange for "indulgences"—the guarantee of a quick, safe passage to heaven. Today, this salvation obsession is evident especially in the Middle East. Martyrs for Islamic terrorist organizations are offered heaven—not just for themselves, but for their parents, brothers, and sisters—in exchange for strapping on a backpack full of dynamite. Meanwhile, in Israel in 2006, a rabbi speaking for the extremely right-wing political party Shas made a television ad in which an angel tells a man that because he voted for Shas, "your place is in heaven." The ad was banned from television.

But if you're in the "grace" camp, how do you think about ethics and personal responsibility? Does God's grace overcome *any* sin? Can murderers go to heaven if they've acknowledged Jesus Christ as their personal savior? (Father Ron Ashmore, who ministered to Oklahoma City bomber Timothy McVeigh in the months before his death, believes they can. In an Easter sermon before McVeigh's execution in 2001, Ashmore said he dreamed of those slain in the bombing welcoming their killer in heaven, saying, "We have been waiting. It is so good to have you home.") Can you be mean to your parents? Can you have sex with your neighbor's wife? Can you wage a war that kills innocent children, say, or fight to deprive people of their civil rights? If you believe that grace is attainable only through Jesus, then do you also believe that those who don't follow Jesus—either by choice or because they were born at the wrong time or in the wrong place—will be barred from a blissful eternity?

Though the story of the Protestant Reformation provides the strongest and most familiar example of the Grace versus Deeds debate, Jews and Muslims have their versions, too. The conflict between Sunni and Shia occurring now in Iraq—which is actually a centuries-long succession struggle—has become an ideological war over the right interpretation of the Islamic tradition, and by extension salvation. Online, Islamic message boards are full of name-calling on both sides, and when a Shiite mosque was bombed in 2009 in sectarian violence in Iraq, a sheikh responded this way: "Let them kill us. . . . We know what they want, and we'll just be patient. But they will all go to hell." On the West Bank, super-religious Jews punish the not-as-religious for violations of modesty codes and capitulations to secular culture. The owner of an electronics store in Beth Shemesh told the Associated Press in 2008 that roving gangs of self-appointed modesty police would regularly invade his shop, smash his MP4 players, and shout, "This store burns souls!"

Among those who are dead-sure that the path they're on leads to salvation, belief in hell as a real place—a hot and stinking pit full of fire—thrives. In America, belief in hell is highest—80 percent or more—among evangelical Christians, members of African American churches, and Muslims, according to a 2008 poll by the Pew Forum on Religion and Public Life. If there's a right way to God, this thinking goes, there's also a wrong way; if heaven is real reward for righteous believers, then hell is real punishment. Among those inclined to see the rules of salvation as unknowable, belief in hell is on the wane. Just 56 percent of mainline Protestants believe in hell. Among Jews, the number is 22 percent.

But as the proponents of irreconcilably opposing dogmas slug it out, another, more benign force is at work. More and more Americans are saying that while they prefer the path they're on, other paths work as well. In its 2008 poll, Pew asked people whether they believed that those of other faiths would attain heaven and a surprising number—70 percent, including 56 percent of evangelical Christians—said yes. This caused a stir, not surprisingly, since exclusivity has always been a bedrock belief of evangelicals. As the articles of faith of the South-ern Baptist Theological Seminary put it, "saving faith" rests on Christ alone "for justification and eternal life."

Pew promptly polled directly on Grace versus Deeds, and the re-sults clearly illustrate that the fundamental conflicts that produced the Protestant Reformation have yet to be resolved. About a third of American Christians believe "being a good person" gets you to heaven, and a third believe that "belief in Jesus" gets you there. About a fifth either don't believe in heaven or don't have a clue. In a 2006 inter-view, even the evangelist Billy Graham, when asked whether heaven would be closed to good Jews, Muslims, Hindus, or Buddhists, gave this surprising reply: "Those are decisions only the Lord will make.

It would be foolish for me to speculate on who will be there and who won't. . . . I believe the love of God is absolute. He said He gave His Son for the whole world, and I think He loves everybody regardless of what label they have." Graham's statement ignited a bonfire online: the most famous Christian evangelist in the world was denounced as an apostate.

In a world that's unfair, where bad people succeed and good people suffer, the notion that what you *do* in life has any relation to where you go after death can seem, at the minimum, a setup for cosmic disappointment. This, at least, was the view of the sixteenth-century Protestant Reformers—led by the German monk Martin Luther and the French layman John Calvin. Disgusted with the orgy of works that characterized the centuries leading up to the Reformation, Luther and Calvin wrote that the best thing—indeed the *only* thing—for a Christian to do was to throw himself on his knees before God and beg for mercy. (When Calvin's Swiss predecessor Huldrych Zwingli lay ill with the plague, he wrote this devastating prayer-poem: "Do as you will / for I lack nothing. I am your vessel / to be restored or destroyed.") Calvin and his theological successors took this idea one step further: even begging for mercy was insufficient. All-powerful God had determined before any individual's birth—even before the creation of the world—who was saved and who was damned. Thus, there was nothing in the world anyone could do about it. This idea, called predestination, continues to live on in American Protestantism, and in certain Jewish and Islamic traditions as well. Jehovah's Witnesses believe that at the end of the world, only 144,000 souls will ascend to heaven to be with God. Everyone else will exist on the newly established earth. The 144,000 (a number in Revelation) are chosen by God; the chosen have no say over the matter themselves. According to certain Muslim hadiths, an angel from Allah sprinkles soil from the place of a man's death in his mother's womb. Even before he is born, God knows where and when he is going to die. During the High Holy Days, Jews chant the *U-netanah tokef* prayer, in which they cede control of their lives to God—who decides "who shall live and who shall die . . . who by sword and who by wild beast . . . who by strangulation and who by stoning."

Instinctively, though, we hope for a little more say in the matter. We want justice for ourselves and for the people we love; we want our enemies to suffer. Bad things happen to good people, we tell ourselves, but in heaven God will fix that. According to the book of Daniel, the way to get to heaven and live among the stars forever is to "lead many to righteousness." The Hebrew word for "rightness" is *tzedek* and it is related, semantically, to the word *justice*. In the Hebrew Bible, a "righteous" man is one who behaves ethically and morally according to God's law. In Daniel's view, "righteous" Jews were those who retained their Jewish practice and belief amid the pressures of mainstream Hellenistic culture. Similarly, the book of Revelation teaches that "righteous" Christians were those who refused to make sacrifices to Roman emperors or to worship other idols. In the Qur'an, Allah promises that "the earth shall be inherited by my righteous servants." In all three holy books, heaven is the place of ultimate justice. The fights over "righteousness," then, amount to this: What does God want from us?

MARTYRS

In all three monotheisms, the most righteous have always been the martyrs; dying for the faith continues to be viewed as a first-class ticket to heaven, a straight shot to the throne of God. I knew the assassinated American journalist Danny Pearl slightly; we worked together at the *Wall Street Journal*. The year after Danny's death at the hands of a group calling itself the National Movement for the Restoration of Pakistani Sovereignty, I was interviewing the Lubavitch rabbi Manis Friedman and asked him whether he believed that Danny was in heaven, since in life Danny had not been a very observant Jew. The rabbi's answer was, unequivocally, yes. In his view, Danny was a martyr: in the moments before his ghastly death, Danny declared (whether voluntarily or by force, we will never know) on videotape, "I am a Jewish American from Encino, California. My father is Jewish, my mother is Jewish, I am Jewish." Rabbi Friedman told me, "That's a ticket to heaven right there."

When Essenes were being tortured and killed by the Romans for refusing to forsake their dietary laws, the first-century Jewish historian Josephus described the scene: "[Not once] did they cringe to their persecutors or shed a tear. Smiling in their agonies, mildly deriding their tormentors, they cheerfully resigned their souls, confident that they would receive them back again." They were going straight to heaven.

In Jerusalem, during the centuries before the birth of Christ, Jews showed their devotion to God through blood sacrifice. "Righteousness" was connected to the ritual slaughter of goats, lambs, and birds at the Temple's altars; if regularly and properly done, such sacrifices bestowed upon the Jews the assurance of God's continued favor. (Many passages in Leviticus are devoted to the right and wrong ways to kill animals.) The earliest Christians understood that the execution of Jesus was the ultimate sacrifice: that is why the Gospels refer to him as "the Lamb of God." With the crucifixion, Christian theology says, God himself—fully human and fully divine—died. He rose again so people could know that death itself was conquered. Going forward, that sacrifice would be the one and eternal symbol of the salvation of all who believe.

Martyrdom, already understood in the Jewish world as a ticket to heaven, became, in the Christian context, a celebrated way to die.

Under persecution by Roman authorities, Christians who suffered death rather than deny their faith were martyrs. The Greek word μάρτυς (martyr) means "witness." They were literally witnesses to their faith. Ignatius, the Bishop of Antioch, was sentenced to death around 107 for defending Christianity to the emperor Trajan. As he was being transported to Rome to be fed to the lions, he wrote letters in which he imagined heaven. "Let fire and the cross; let the crowds of wild beasts; let tearings, breaking, and dislocations of bones; let cutting off of members; let shatterings of the whole body; and let all the dreadful torments of the devil come upon me: only let me attain to Jesus Christ." The lions devoured Ignatius so thoroughly that only the merest scraps were left. In the early centuries of Christianity, the stories of the martyrs were so popular, and such powerful tools of evangelism, that the Church Father Tertullian wrote, "The blood of the martyrs is the seed of the church."

Therefore, when Constantine converted to Christianity—and, in 313, finally gave the religion legal recognition throughout the empire—Christians had to rethink their identity. No longer a persecuted minority, they could not emulate Christ literally by turning themselves into blood sacrifices. Instead, they linked themselves to his martyrdom through liturgy and worship; by partaking of the sacramental wine and bread of Communion, the earliest Christians were connecting themselves to the blood their Lord shed. (Holy Communion refers, of course, to the Last Supper—the Passover meal that Jesus ate with his disciples on the eve of his execution. The Gospels contain the words that every Christian knows by heart: "While they were eating, Jesus took a loaf of bread, and after blessing it he broke it, gave it to the disciples, and said, 'Take, eat; this is my body.' Then he took a cup, and after giving thanks he gave it to them, saying, 'Drink from it, all of you; for this is my blood of the covenant, which is poured out for many for the forgiveness of sins.' " Jesus goes on to say that he will not drink wine again until he reunites with his disciples in paradise.) But some Christian communities also connected themselves to Jesus's life, death, and resurrection through asceticism, celibacy, poverty, and constant prayer. Monks became the latter-day equivalents of the martyrs.

Around 530, an Italian monk named Benedict wrote down his Rule, a guide still in use by Roman Catholic monasteries all over the world. Benedict's Rule dictates monastic life down to the tiniest detail—how many times a day to pray, what to wear, how much to eat, when to laugh; it particularly instructs monks in their view of death, God's judgment, and heaven. "Yearn for everlasting life with holy desire," Benedict wrote. "Day by day remind yourself that you are going to die. Hour by hour keep careful watch over all you do, aware that God's gaze is upon you, wherever you may be." In Benedict's view, monks must aim to live at all times as if they already inhabited heaven.

Father Dominic Whedbee is not a Benedictine monk—though the order still exists—but a Trappist, a member of the strict Cistercian order, established in eleventh-century France to get back to the basics of Benedict's Rule. As Father Dominic explains it, monks are ordinary humans attempting to live according to the teachings of Christ, and their job is to pray constantly for the salvation of all the souls of the world. With their communities, their songs, their costumes, the design of their dwellings, they consciously try to create a mirror of heaven. Their prayers lift everybody in the world up to God.

I met Father Dominic on a perfect July day, at St. Joseph's Abbey, in Spencer, Massachusetts. The monastery is situated on a hilltop, which, when the sky is clear, affords views of Connecticut, Massachusetts, and New Hampshire. The low stone buildings seem to have grown organically up from the ground; on the day I visited, the grass

ruffled, the air vibrated with the sound of far-off gas mowers and ci-
cadas, and elderly tourists perambulated hand in hand along winding
stone pathways. Fruit trees flowered. Through a window, I saw a white-
robed monk playing the harp.

Father Dominic is sixty years old, but—I am reminded of Saint
Anthony's angelic appearance—his face is as unlined as that of a man
thirty years younger. His hair is still blond, his eyes clear. He wears a
long white robe, a brown leather belt, and sandals. He talks more or
less incessantly, which is not surprising since he's lived at the monastery for twenty-six years among seventy brothers whose vows prohibit
them—except on infrequent, prescribed occasions—from speaking.
They are not allowed to own any property. They see their families only
four days a year. The primary sound they make with their voices, day
after day, year after year, is the sound of the Divine Office—psalms
they chant seven times a day, until they briefly sleep and wake to
chant again. As the public face of the abbey, Father Dominic had
permission from his abbot to speak to me.

The Cistercian order was established in reaction to what its founders
saw as the corruption and materialism of medieval monasticism. One
of its most famous members was Bernard of Clairvaux, who avidly encouraged Christian men to sacrifice themselves—literally, on the bat-
tlefield, or figuratively, in monasteries—to attain heaven. "Clothe not
yourself in sackcloth," he preached to the soldiers at the beginning of
the Second Crusade in 1146, "but cover yourself in your impenetrable
bucklers. . . . Hasten to expiate your sins by victories over the infidels
and let the deliverance of holy places be the reward of your repen-
tance. . . . Abandon then the things that perish to gather unfading
palms and conquer a Kingdom which has no end." Give the blood in
your bodies to Christ, in other words, and you will gain heaven.

The Cistercians were thought to have special favor in heaven, and in the late Middle Ages, people would pay them to say auxiliary prayers for their dead, hoping these incantations would launch the dearly departed more quickly through purgatory to paradise. Although they no longer accept payment for their prayers, the monks of St. Joseph's still pray for the dead all the time: at the end of every office, at the end of every meal, and during mass.

The rigors of life at St. Joseph's are relentless, and Father Dominic is full of rueful good humor about them. He and his brothers eat only vegetables. ("I never tasted a rutabaga before I came here.") They share everything, even their underwear and T-shirts. Every action of every day is prescribed and choreographed. There is a right way to kneel in the chapel, to nod hello to a monk passing in the hall, to pluck a library book off a low shelf. When not in prayer, they work; some make the Trappist jams and jellies they sell nationwide, others manufacture vestments and altar cloths. Dominic speaks exuberantly of the "party" they have the day after Christmas, which most of us would simply call lunch: they haul out their instruments, play chamber music, sing carols, and eat cream cheese sandwiches, potato chips, and pickles. When a member of the group is sick or dying, his brothers do what they can to care for him without outside intervention. The brothers who lie beneath the forty-seven wooden crosses in the graveyard at the center of the monastery are as much a part of the community as the living—and with vocations shrinking, the dead will soon outnumber the living. "Mortality isn't abstract here," Dominic tells me.

The object of this radical experiment in communal living is—and has always been—not just for each individual member to assure himself a place in heaven, but through constant prayer to gain heaven for the whole community, and even for all the souls in the world. Dominic, who grew up the son of a Maryland doctor and who dreamed himself of becoming a doctor and a family man, describes the culture shock he first experienced here. "Suddenly, I'm living with these people—and they don't speak. Getting up at two a.m. gets pretty old." Like any intimate group, the brothers squabble and condescend to one another; they use their brief chances at conversation to make cutting remarks; they tattle on one another to the abbot. But Dominic, unlike many other religious I've met, does not seem broken, or even cracked; he is wry and self-knowing. Even in these circumstances, he seems like a normal guy.

Finally, eventually, Dominic submitted to the task before him: face God, pray to God, constantly, every minute, on behalf of humanity. "I realized, 'You're playing for keeps. . . . What else is there to do here? Make jelly? Please." When young monks come to him, at the point of giving up, he urges them to "keep working on it and don't get discouraged—that's letting eternity crack you open."

Here is what affects me most about Dominic's sacrifice. He believes that all of us—me included—are a whole, and that unless we all go to heaven, no one will. Twenty-six years into his peculiar way of life, Dominic still mourns what he's lost on earth: a profitable career, a family of his own, even the chance to be a doting uncle to his sister's children. It feels crazy to be saying this, but Dominic's martyrdom consoles me. I am too distracted with my job, my family, my commute, and my grocery list to pay real attention to the fate of my soul. Dominic's silent life of prayer gives me comfort and hope. I actually believe that if anyone's prayers can get me to heaven, Dominic's can.

GRACE VERSUS DEEDS

But if you couldn't be a monk or a martyr, then at least you could do the right thing, and as early as the fourth century, Christians were making lists of sins and their correlated remedies. Do this, then do that, and salvation is yours. The fourth-century Hellenistic monk Evagrius of Pontus established eight deadly sins, which, two hundred years later, Pope Gregory the Great edited down to the familiar seven: vainglory, envy, anger, melancholy, avarice, gluttony, and lust. Pride, he said, was the root of them all. In early Christian communities, penance was public: On Ash Wednesday, at the beginning of Lent, people who had committed serious sins, such as idolatry or murder ("mortal" and "venial" were not yet fine-tuned terms of art), would become "penitents." They wore special clothes, sat together in church, and were barred from taking communion. If single, they were expected to remain unmarried; if married, they were expected to leave their families. So severe were these requirements that many sinners did not become penitents until their lives were nearly over. In the sixth century, Caesarius of Arles complained that too many sinners were also procrastinators: "People say, 'When I am old I shall undertake penance. When I am grown old or desperately ill, then I shall ask for penance.' "

In the fifth and sixth centuries, certain bishops began to publish "penitentials"—handbooks or ledgers of sins and corresponding punishments. These circulated widely, especially in Europe and the British Isles. The educated clergy assumed that "Christ wants everyone to be saved, he died for all of us," explains Jeffrey Burton Russell, emeritus professor at the University of California at Santa Barbara, who has written three books on heaven. "But is stealing a loaf of bread equal to murdering your mother-in-law?" In their efforts to sort and weigh sins, the authors of the penitentials were detailed and discriminating: a person who murdered a cleric, for instance, had to do more penance than one who murdered a sibling. Incest, sodomy, bestiality, and adultery earned penances, of course, as did priests and nuns who kissed— or worse. Penalties could be levied upon those who ate horsemeat, made amulets, drank magic potions, and raised the dead.

Penance was not for sissies. The most common penalty was prayer and fasting (which sometimes reputedly resulted in death), but other methods of punishment—such as sleeping in water, on nettles, on nutshells, or with a corpse in a grave—were possible as well. Some bishops ordered penitents to assume physically stressful positions: they had to stand with arms outstretched while singing praises to God, or pray with their hands stretched upward but body bent. The mid-seventh-century Irish Penitential of Cummean metes out penance for every kind of bad or rude behavior, including vomiting on your Communion wafer (forty days' penance), as well as eating scabs, lice, or one's own "excreta." The penance for the last: "an entire year on bread and water."

Thomas Aquinas (1225–1274) is credited more than anyone with codifying the idea that specific actions on earth correlate to a specific place in the afterworld. Aquinas made distinctions between mortal sins—which, without confession and absolution, permanently separated people from God and prevented their entry to heaven—and venial sins, which did not. He imagined a stratified and hierarchical heaven: although everyone there is content, some people are seated closer to the Throne than others. "The more love [of God] someone will have [in heaven], the more perfectly one will see God, and the more blessed one will be," he wrote in his *Summa Theologica*. "All the blessed see the Highest Truth but they do so in various degrees." God gave salvation to the world through Jesus, Aquinas taught. But he also gave people free will. It was through the use of that will that individuals would attain salvation—or not.

In the Jewish world, at around the time of Aquinas, the physician and philosopher Moses Maimonides (1135–1204) was also busy sorting and weighing sins. Jewish tradition had long held that God commanded Jews to follow 613 rules, or mitzvoth, which dictate every aspect of Jewish life, from prayer—how often and when—to dietary rules. (In the Jewish mystical tradition, each Hebrew letter has a numerical value. The number 613 thus is derived from adding up the values of the letters in the word *Torah*—plus two for the commandments that predate the Torah: "I am the Lord your God" and "You shall have no other God before me.") But what were the rules, exactly? The rabbis disputed and reformulated their lists endlessly. Around 1168, while preparing to write his masterwork, the Mishneh Torah, Maimonides attempted to codify, once and for all, the 613. At the same time, Maimonides also came up with a list of thirteen general principles, which have become so central to Jewish worship that the Orthodox still recite them each morning in a prayer called Yigdal. They include the belief in One God, as well as the belief that God has no body, that the Torah comes from God, that the Messiah will come, and that the human body will be resurrected. Any Jew who disputes these principles will, Maimonides said, forfeit their stake in the hereafter. "He said very sharply, more sharply than others," says David Berger, a Maimonides scholar at Yeshiva University, "that if you do not believe, you do not have a portion in the world to come."

Over breakfast at a kosher restaurant in Queens, Berger explains that Maimonides allows for a view of salvation that includes what Christians would call grace *and* deeds. Jews have always believed that God watches what they do and keeps a tally of their mitzvoth. At death, the sum of a person's mitzvoth determines how high he or she ascends to God and how long he or she must endure purgatory (more on this soon). But Maimonides also held that a person's goodness—or righteousness, if you will—"is measured by the *weight* of the good deeds, and not by the number," Berger says. And humans, who have no way of knowing which deeds in the final tally will carry the most weight, must continue to endeavor to please God through charity and good works. Although Berger didn't put it quite this way, God's supernatural accounting system is the Jewish version of grace. Modern Jews are pleading for God's grace when they say this prayer on the High Holy Days: "We have no deeds. Do unto us charity and loving-kindness and save us." It is perhaps worth noting here that Maimonides lived most of his life in Muslim Spain and, eventually, Egypt—even working as the personal physician to the sultan Saladin. Scholars continue to debate the extent to which Islam influenced his philosophy (and, indirectly, every Christian and Jewish theologian he subsequently influenced), but there is no doubt that it did.

Hadia Mubarak is a young Muslim American activist and a graduate student at Georgetown University. She talks about getting to heaven much the way Maimonides did. In Islam, she explains, attaining paradise "is the motive behind everything you do." At the same time, "you don't know which action will take you to heaven. There's no specific value attached to any act of charity."

Mubarak decided to put on the *hijab*, the traditional head covering of religious Muslim women, when she was twelve years old. She attended a Muslim junior high in Panama City, Florida, and like many girls in her circle, started covering her head when she reached puberty. "It was," she told me, "a natural thing to do." Her decision evoked no ire from her parents. Both her mother, who was born in Jordan, and her elder sister cover their hair. Not all her friends had such an easy time— casting off the *hijab* was, among some in her mother's generation, a sign of Western liberation and freedom. "Some of my really close friends decided to put on the headscarf after they went to college and their parents were really upset," she said. "They felt like their children were throwing away all the opportunities their parents worked so hard to provide . . . they felt that people would judge them based on the headscarf and they'd never stand a fair chance in the career world."

Now twenty-seven years old, Mubarak is married and the mother of
a two-year-old son. She was the first female president of the national
Muslim Student Association, which has chapters on campuses across
the country—many of which are known for their outspoken, anti-Israel
positions. I heard her speak once at Georgetown, and was astonished at
her rhetorical gifts. She speaks with the clarity and passion of a much
older person. Mubarak is outraged over the killing of Palestinians by
Israel and infuriated at the way American Muslims were treated by the
media and their neighbors in the years after 9/11. She will say—and
has said—this bravely to anyone who asks. And yet, her appearance
tells a different story. Petite, dark-eyed, shrouded, and smiling, wearing a skirt down to her shoes and a sweater that hides her body, she
is the picture of obedience and submission. In Mubarak, conventions
of womanhood, religious observance, and cultural and ethnic identity
are being radically redefined.

We met at a Georgetown coffee shop to talk about heaven. Mubarak was squeezing me in between classes. The shop was too crowded,
we hadn't much time, but in spite of the bustle, Mubarak appeared
calm. We launched into our topic without any small talk, and almost
immediately Mubarak was talking about the Grace versus Works
debate. "Who's going to heaven? The man who prays five times a day
his whole life? Or the one who helps a lady across the street? This gets
repeated a lot in Friday sermons: no matter how good you are, you
can't earn heaven."

Still, she does everything she can to be heavenbound. She wears
the *hijab*, she prays five times daily, she gives to charity. She does not
drink alcohol, as the Qur'an mandates; she does not take out interest-
bearing loans. She tells me the story from the hadith of the righteous
man who went to heaven. He accrued a mountain of good deeds: he
prayed and gave to charity every day of his life. And yet, when the

angels weighed his deeds against a single one of God's gifts to him—
his eyesight—the balance tipped overwhelmingly in God's favor. That
is God's beneficence.

If you can't earn your place in heaven, I asked her later, then why try
so hard? Why not leave your fate up to God? "Works definitely matter
and God makes that clear in the Qur'an," Mubarak responds by e-mail.
"The idea of God's mercy is that in comparison to everything God has
given us, nothing we do in this world can measure up or compensate.
Hence we hope to enter heaven through God's mercy, but we hope to
receive God's mercy by our work and faith. Does that make sense?"

In the afterlife, Mubarak believes God will reward her for her piety.
"No one enjoys living a restrictive life," she tells me. "I know I'd look
a lot better if I wasn't wearing any scarf and dressed a certain way. Islam's modesty requirements are a huge sacrifice for any woman. But it's
to please God. It's what God wants in the end. In the end he'll reward
us with not having any restrictions."

PURGATORY

By the twelfth century, the Roman Catholic hierarchy—more than
any Muslim or Jewish group—had established the idea that sins could
be counted, weighed, discounted, and paid off in life. The first step
was baptism. Even the earliest Christians understood baptism as a rite
that would remove original sin, the stain of Adam, from their souls
and open the door to heaven. Babies were thus to be baptized as soon
as possible. The question then almost instantly arose: What about
babies who died before baptism? What of their souls?

The early Church Fathers approached the question with finesse.
The fourth-century bishop Saint Gregory of Nazianzus argued that
unbaptized children "will be neither glorified nor punished by the

righteous Judge, as unsealed and yet not wicked, but persons who have suffered rather than done wrong. . . . For not every one who is not bad enough to be punished is good enough to be honored." They would go, in other words, to an in-between place, which today we would call "Limbo." Augustine was harsh in response. Unbaptized babies had original sin. They would go to hell. In 418, he convinced the Council of Carthage to condemn the notion of "an intermediate place, or of anyplace anywhere at all, in which children who pass out of this life unbaptized live in happiness." This hell for babies would not be so bad, Augustine added, softening the blow. The punishments would be mild; the babies' souls would barely notice.

Still, condemning newborn babies to hell discomfited many, and in the twelfth century the French philosopher and theologian Peter Abelard argued that the babies in Limbo suffered no physical torment at all, only a spiritual separation from God. It was Thomas Aquinas, finally, who quashed the idea that unbaptized babies merited any kind of punishment. Limbo, he said, was a happy place. It was not heaven, not in God's actual realm—"a second-class heaven," as Father Thomas Reese, a research fellow at the Woodstock Theological Seminary, puts it—but no matter. The babies there were so attuned to God's plan that they didn't know what they were missing.

This was the idea that so many of my friends who went to Catholic school learned at the hands of the nuns. Limbo was a place for unbaptized babies, apart from God, but contented nonetheless. Cartoon renderings showed winged babies floating around in featureless ether. (I know many a Catholic mother who, having lost a baby through miscarriage or at birth, adamantly refused to accept that teaching. "My baby is in heaven," a friend told me. "I know she is.") In 2007, Vatican scholars under the direction of Pope Benedict XVI issued a forty-one-page report in which they encouraged Catholic educators to phase out

Limbo. It was a welcome decision. "People felt uncomfortable about babies who weren't baptized being thrown into this waiting area," says Reese, "and they felt that people who had lived good lives or hadn't sinned—there was no reason they should be kept out of heaven." A similar sentiment led the early Protestant Reformers to edge away from the idea that infant baptism was a prerequisite for salvation. "God declares that he adopts our babies as his own before they are born," wrote John Calvin, the father of predestination, in 1536. These days, most Protestant denominations teach that unbaptized babies go directly to heaven, obviating the need for Limbo.

Purgatory is different from Limbo. In the Middle Ages, baptism was just the first step on the path to heaven. Public penance had given way in most places to private confession, which was necessary for the expiation of sin. Cleansing the soul to prepare it for heaven became an obsession of lay Catholics, who learned in church that their deeds—and misdeeds—on earth would correlate to their place in heaven. In *The Canterbury Tales*, written in 1380, Geoffrey Chaucer asks God's forgiveness for anything in his work that might cause offense, including "many a song and many a lecherous lay." "Grant me the grace of true penitence, confession and expiation in this present life," he writes, "through the benign grace of Him who is King of kings." The scholastics, Catholic intellectuals in the great European universities who, inspired by the Greeks, aimed to develop a rational theology, refined the getting-to-heaven process by renovating a concept that had long existed in monotheistic tradition: purgatory.

By 200 CE, the rabbis had already begun to formulate an idea of an "in between" place, where souls who were neither completely good nor completely bad would abide until they were pure enough to ascend to heaven to be with God. A rabbinic treatise from the late first century CE refers to three kinds of souls: "the truly holy," "the truly wicked," and those "in between." "The third group shall go down to Gehenna [hell] for a time and then come up again, as it is written." The Kaddish prayer, widely in use as a mourners' prayer by 1200, helped a soul move through the cleansing process and up to heaven. Jewish tradition teaches that close relatives should say Kaddish for twelve months, but most mourners do it for eleven; only the worst offenders require an entire year of prayer. Afterward, even the most tainted souls are ready for God. Few, if any, are barred from heaven completely. (Through friends I met Aric Press one Sunday morning in Brooklyn. He is a Jew who goes to a Conservative synagogue, and tells the story of the months in 1998 when he said Kaddish for his mother. Never having said Kaddish before, he did not know when to stop. A rabbi advised him to stop a week shy of the designated eleven months, "because," as Aric remembers it, "if it took the whole eleven months, people would think she was *really* bad.")

Islam also talks about an "in between" place (or time) called *bar-zakh*—wedged not between heaven and hell, but between death and resurrection. According to Islamic tradition, two inquisitor angels—black with green eyes, in some accounts, and with supernaturally long fangs—enter a person's grave after death. They are terrifying. They wake the soul and ask it questions to determine its degree of faithfulness. Based on the answers, the angels decide where that soul will abide to await the final Judgment. According to the eleventh-century Persian theologian Abu Hamid al-Ghazali, souls fall into four categories: The best people are shown from the grave a window to paradise. The second-best are shown a window to hell and then reassured—they won't go there. The third, who fail to answer their questioners with sufficient clarity, are beaten, their graves set on fire. The last group, the profligates, are visited in the grave by their evil deeds come to life as wild animals.

Tradition holds that after this grueling question-and-answer period, Muslims sleep until the Judgment, but here there are variations. Al-Ghazali suggests that the damned wander the earth as restless spirits, while the blessed enter paradise after a short, perhaps months-long, wait. John Voll, a professor of Islamic history at Georgetown, reminds me in a conversation in his office one snowy afternoon that our own folk literature is full of stories about sleeping—and waiting—and awakening to find everything changed—a popular genre in the medieval Middle East among all religious groups. "Rip Van Winkle" and the movie *Awakenings*—the true story of hospital patients paralyzed by Parkinson's disease who experience, briefly, the sensuous vitality of normal life—are variations on this theme. In these stories, "heaven" is waking up from a long sleep to find everything more real, more vibrant than it was before.

Clearly, then, the idea of a waiting place—or perhaps it should be called a process—that occupied the space between death and heaven had existed for hundreds of years in monotheistic religion. By the twelfth century, interest and belief in a Christian purgatory had reached a peak. Peter the Chanter, a French theologian, reflected the popular understanding: "the good go either at once to Paradise if they have nothing with them to burn, or they go first to Purgatory and then to Paradise, as in the case of those who bring venial [that is, forgivable] sins with them. No special receptacle is set aside for the wicked who, it is said, go immediately to hell." Purgatory, according to Peter the Chanter, was a place: the amount of time you spent there—and the horrors you experienced—depended on the severity of your sins, and the degree to which they had been expiated in this life.

The medieval Christian purgatory was vivid and a lot like hell—the only difference being that purgatory's tortures were finite. Some say the souls in purgatory "are so replete with suffering," wrote Thomas

Aquinas in the *Summa Theologica*, "that they know not that they will
be set free." But that's not the case, he added: the dead in purgatory
require the prayers of those on earth. Tales of purgatory, written down
and passed around among neighbors and in church congregations,
worked as entertainment and as moral deterrents. The tortures described therein were so thoroughly terrifying that they would move
any lax Christian to follow a straighter path. One of the most popular
in the twelfth century was "The Legend of the Purgatory of St. Patrick." In it, an Irish knight named Owein sees enough in a vision of
purgatory to give his life to God.

According to legend, there was a hole in the ground on Station
Island, located in the middle of a large lake in Ireland; when Saint
Patrick was converting the Irish, he used it as a conversion tool. He
would install a pagan overnight in the hole. There, the unfortunate
captive would endure visions of the horrors of purgatory. In the morning, the pagan would either be dead—or a faithful Christian. "The
Legend of the Purgatory of St. Patrick" describes poor Owein's visions
of purgatory during his stay underground. He sees people nailed with
flaming pegs to the ground, baked in ovens, turned on spits, dunked
into molten metal, and strung up on iron hooks. Toads and snakes eat
people up. Finally, the knight calls on God and at last finds himself
on a bridge over a river of fire. He sees marvels, including the earthly
and the celestial paradise. Having survived his vision, Owein makes
a penitential pilgrimage to Jerusalem, and helps to found an abbey
there. The hole where Owein allegedly spent the night was sealed in
1790, and a chapel was built atop it. Today, St. Patrick's Purgatory at-
tracts up to thirty thousand visitors a year.

No wonder, then, that Christians would do anything to shorten or
evade their tenure in purgatory. By the end of the eleventh century, the
popes were granting indulgences—official abridgments of future stays

in purgatory—in exchange for cash payments. In 1095, Pope Urban II
bestowed indulgences upon men who went on a Crusade to Anato-
lia, which is modern-day Turkey, and in 1215, at the Fourth Lateran
Council, the bishops reaffirmed Urban's promise: men who went on a
Crusade themselves (that is, did not pay someone else to go instead)
and at their own expense would have their slates wiped free of all sins
and ascend—like the righteous martyrs—immediately to heaven. In
1300, Pope Boniface gave indulgences to anyone who traveled not to
Jerusalem but to Rome on a pilgrimage and lived to tell about it; he
then extended the benefit to anyone who died on that journey.

Not only did people believe that payments to the Church would
hasten their arrival in heaven; they also believed that special prayers—
and prayers by special people—would, too. Fifteenth- and sixteenth-
century British wills are full of detailed instructions to family members
and local clerics: how, where, when, and how often to pray for the soul
of the deceased. Say "diriges" (prayers for the dead) "as hastily as possible . . . after my departing frome this world" instructed one man or "as
sone as I am deade w'toute eny tarrying," wrote another. Offer trentals
(a series of masses held on thirty consecutive days) "to be doen for me
from the houre of my dethe unto the tyme of my burial," commanded
a third. The prayers of the poor were thought to be more beneficial
than the prayers of the rich. At the most ostentatious funerals, the
poor—shrouded and holding candles—would encircle the corpse at
the mass.

Catholic salvation theology was not so very different then than it is
now. Catholics believe that God's mysterious grace—bestowed upon a
sinner through a confession of faith—opens the door to heaven. But
salvation comes from that *plus* a life of faithful action. The Church
was—and is—seen as both the conduit for God's love in the world
and a kind of intermediary institution, like a bank, to which sinners

make payments in the form of prayers and penance—and receive credit in the afterlife as indulgences. In the centuries preceding the Reformation, the Church's role as banker clearly outweighed its role as a purveyor of grace. Medieval Catholics endeavored to calculate and control their eternal destiny much the way day traders in the boom economy obsessively hedged their online stock portfolios.

Purgatory, then, can be said to have brought about the Protestant Reformation. It was the abuse of indulgences, as well as the mechanical way in which Catholics imagined they could affect their own salvation, that infuriated Luther and Calvin and moved them to reinvent the idea of what it meant to be a Christian. They threw away the idea of the institutional church as a banker that calculated an individual's credits and debits in heaven. All a Christian needed to be saved, they said, was a belief in Jesus.

But belief in purgatory endures. The Second Vatican Council of the 1960s supported it, and among Catholics—especially those nostalgic for old-school, Latin-mass, fish-on-Fridays Catholicism—prayers for the souls in purgatory remain vitally important. According to Catholic folk tradition, a certain prayer to Saint Gertrude releases a thousand souls from purgatory at a go, and the Mission to Empty Purgatory, a Georgia-based ministry, is trying to collect enough prayers to release the 106 billion souls it calculates are potentially confined there. A number of American dioceses, including Brooklyn and Jackson, Mississippi, have revived the practice of indulgences in an effort to bring stray Catholics back home to the church. "Indulgences are a way of reminding people of the importance of penance," said Father Thomas Reese. "The good news is we're not selling them anymore."

Mormons tend to say "worthy" rather than "righteous," but it means the same thing. Are they good enough in the eyes of God to ascend after death to the highest level of heaven? Members of the Church of Jesus Christ of Latter-day Saints—established in America in 1830 by Joseph Smith, after he had a series of visions—believe that all human life is a trial, a test of worthiness. On Judgment Day, they will be able to recall everything they ever did. Should God deem them worthy, they will go to live with Him and become like gods themselves. In LDS theology, all people go to one of three heavens: the Celestial, "whose glory is that of the sun, even the glory of God," according to the Doctrine and Covenants, part of the Mormon canon; the Terrestrial, which is like the moon; and the Telestial, the lowest level. Those people who are most like God, and whose choices in life reflect their divine spirit, will ascend to the Celestial realm. (Faithful Mormons fall into this category.) "But even if you're a schmuck who just wants to watch football, or made your life by kicking people to the curb, you still resurrect to some kind of glory," explains Kathleen Flake, who is a religion professor at Vanderbilt University.

From childhood, Margaret Toscano tried to do everything right. She was born and raised in the LDS church—she is a sixth-generation Mormon. As a little girl, she went to church almost all day on Sunday and to a youth group one night a week. On Monday nights, her father would lead Family Home Evenings, which are traditional in Mormon families. He would read from Scripture (the Latter-day Saints read both the Bible and the Book of Mormon) and ask his eight children to discuss it. Starting around age ten, she, like most Mormon children, would do "proxy baptisms" at the temple. She would fall backward into a big tank of water as an officiator read aloud the name of a dead person, thus allowing the deceased (who was not a Mormon) into the celestial heaven. At twenty-one, Toscano "received her endowments" of the Church—she participated in the secret rites that bind an individual to God. She was anointed with water and oil, and given special undergarments to wear at all times and white robes for special temple ceremonies.

Every two years, Toscano renewed her "Temple Recommend," an official certificate like a driver's license, which testified to her worthiness. To obtain this document she, like every other Mormon, had to submit to an interview with the local bishop, in which he asked questions concerning her sexual purity (chastity when unmarried, fidelity when married), tithing, adherence to dietary laws (Mormons don't smoke, drink alcohol, or drink coffee or tea), honesty, and obedience to church elders.

Toscano passed each test, and she loved her religion, but she felt depressed, hemmed in. She married her first husband at a civil ceremony, and divorced him two years later. As a young woman, she told me, as we sat together on a bench in a public park in Salt Lake City, she was often full of self-loathing, unable to shake the feeling that she was falling short. At twenty-nine, she remarried—this time to a hardworking Mormon lawyer, a convert from Catholicism. The ceremony took place in the temple. Together, she and her husband were "sealed." They could now expect to live together forever in heaven—and any children they had would be part of that divine family unit, too.

But Toscano and her husband, Paul, began to chafe against the constraints of their religion. They had four daughters, and Margaret put herself through graduate school in classics and began to study feminism. Together with half a dozen other Mormon intellectuals, Paul began to argue against the authoritarianism of the church. He was excommunicated in 1993. Church officials took notice of Margaret around that same time, when she wrote a series of scholarly papers arguing that Joseph Smith never intended Mormon women to be excluded from the priesthood—that they should be allowed to ascend to the highest levels of authority and power in the church.

On November 30, 2000, when she was forty-six years old, the authorities held a hearing and asked Toscano to repudiate her own work. She would not and she was, officially, excommunicated—though for some years already she had been slipping away. She had stopped wearing the sacred undergarments and even drank the occasional glass of wine. "Taking that first glass of wine," she remembers, "was like being polluted. There was a way in which, as an LDS member, I used to feel purer than other people."

Heaven still matters to Toscano because it matters so much to her family members within the Church. According to the tradition in which she was raised, families stay together in heaven. But since she was cut off, Toscano is no longer eligible to attain the Celestial Kingdom, which means, according to Mormon doctrine, that she will not be with two of her three sisters—one living and one dead—in the highest level of heaven. (The third, excommunicated as she is, will join her in a lower level.) In her heart, Toscano does not really believe this teaching—"I don't believe God would separate people who love each other," she says—but her childhood training is hard to dismiss and together, she and her sisters fret, half-seriously, about their places in eternity. Margaret and her devout sister joke about heaven—because if they don't, they'll cry. Perhaps there's a Plexiglas wall between the levels of heaven, Toscano has said to her sister, like the kind you find in prisons. Perhaps we'll be able to see each other, and talk on the phone.

"You live all these rules and it forces you to grace, because there's no way you can ever feel good enough inside," Toscano says, invoking one of her heroes, Martin Luther. The "maddening legalism" of the LDS church "drove me to grace," she says. She has not attended any church since she left the Latter-day Saints.

REFORMERS

"If anyone could have gained heaven as a monk then I would indeed have been among them. . . . I lost hold of Christ the Savior and comforter and made him a stock-master and hangman over my poor soul." These agonies were written in the 1500s by Martin Luther, about his years as a monk in an order called Observant Augustinians. He tried, he said, to obey the rules. He fasted. He flagellated himself. But nothing in the rules was moving him closer to God; moreover, the widespread abuse of the rules he saw around him disgusted him.

In the fifteenth century, the popes had begun granting indulgences to rich families in exchange for cash payments and other lavish gifts, and in 1517 Pope Leo X held a fund-raiser to rebuild St. Peter's Basilica. He hired a Dominican monk named Johann Tetzel to tour Europe, offering indulgences for cash. Tetzel, a born marketer, had come up with this slogan: *Sobald das Geld im Kasten klingt / Die Seele aus dem Fegfeuer springt!* ("As soon as the gold in the casket rings / The rescued soul to heaven springs!") Tetzel carried with him an inventory of sins and the payment required to expiate them. He reportedly claimed his indulgences could even rescue someone who had raped the Virgin Mary.

Luther preached against Tetzel, and Tetzel, hearing of Luther's sermons, warned that the monk was a heretic. On October 31, 1517, Luther nailed his ninety-five theses—complaints against the Roman Catholic hierarchy—to the church door in Wittenberg, Germany. The last two items related expressly to heaven: "Christians are to be exhorted that they be diligent in following Christ . . . and thus be confident of entering into heaven." All you have to do, said Luther, is believe. Christians no longer needed to count and weigh their sins and pay or pray for absolution. In Luther's—and his fellow Reformer John

Calvin's—view, heaven was not a stratified, hierarchical place, where some fared better than others, or where any authority existed outside of God. "We will be equal to St. Paul, St. Peter, our beloved Lady, and all the saints in their honor and glory," wrote Luther.

The Protestant Reformation gave heaven back to all the saved on equal terms, but it also arguably made the place a lot less fun. Its image—each soul communing with God, an eternity of praise and song—pales in comparison to the opulent, glittering, and crowded Catholic version. And though in the years following the Reformation, Puritan ministers did what they could to keep their followers from populating heaven with friendly saints and fellow pilgrims—sometimes even whitewashing the church frescoes that depicted such scenes—it was not too long before, in America at least, people began to want something more.

SEVEN

VISIONARIES

Don Piper was driving his red Ford Escort from a meeting to his church near Houston on a rainy January day in 1989. There was a thick fog, and Piper had just slowed to read a plaque on the side of the narrow bridge across which he was driving when an eighteen-wheeler crossed the center line and crushed his car. When they arrived at the scene of the accident, the emergency medical team found Piper lifeless. His left arm dangled behind him. His left leg was shattered. The medics pronounced him dead, put a tarp over the vehicle, and proceeded to assist passengers in other cars. That's when Piper, who was thirty-eight years old, a father of three, and a Baptist minister, went to heaven.

The story he tells—in his book *90 Minutes in Heaven* and in person as he travels the country—is both familiar and moving. He remembers an enveloping light and a feeling of intense joy. He remembers standing before an ornate gate and being greeted by a crowd of familiar faces—his grandfather, a high-school friend, his great-grandmother who in life had been toothless but now had perfect teeth. As he recounts his experience, he repeatedly says that although he knows he must tell his story chronologically, time has no meaning in heaven. He also insists, even as he tries to describe them, that the sights he saw there were indescribable. "Everything I experienced was like a first-class buffet for the senses," he writes. He saw light, which grew brighter and brighter but was never blinding. "I was amazed that the luster and intensity continually increased," he writes. "The farther I walked, the brighter the light. The light engulfed me and I had the sense that I was being ushered into the presence of God. Although our earthly eyes must gradually adjust to light or darkness, my heavenly eyes saw with absolute ease."

And then he heard the music. At first, it sounded like the swishing of birds' wings—familiar from his boyhood in Arkansas. He soon realized it was the wings of angels. The sounds of heaven "differed from anything I had ever heard or expect to hear on earth"—as though three different CDs were playing at once, he says, yet all the sounds blended together into one jubilant noise. Even now, when Piper lies down to sleep, he can hear the sounds of heaven.

Piper was brought back, he says, by a colleague who happened upon the scene, and, following a command from God, climbed into the back of the Escort to pray. When the friend started to sing a hymn—"What a Friend We Have in Jesus"—Piper joined in. The friend rushed over to the EMTs, screaming, "He's alive!"

Piper had thirty-four surgeries. His recovery was slow, his wounds more than physical. He suffered anger, depression, and suicidal thoughts. Finally—through long conversations with understanding friends—he came to understand that his personal catastrophe was a message from God, and he devoted his life to spreading the news: heaven is real. He travels constantly—Georgia, Texas, California, Florida, and Alabama in a single month—speaking mostly to church groups and at retirement communities. His pain is constant; he always asks hosts to put him up on the ground floor so he doesn't have to climb stairs. "People want to know that God does answer prayers," he told me. "They want to know that miracles can still happen." Published in 2005, 90 Minutes *in Heaven* was on the best-seller list for more than two years; a sequel, *Heaven Is Real,* came out in 2007.

Is Don Piper a crackpot, a huckster—or a prophet? The same has been asked of every visionary since long before Jesus. Piper falls into a long tradition of travelers to heaven—men and women who say they find themselves, usually by accident, in the Kingdom of God and return, changed, to speak truth to a skeptical world. The vision of Jacob, the books of Enoch and Revelation are the early monotheistic precursors, but of course there are others. The Mesopotamian legend of Gilgamesh, written about four thousand years ago, describes the journey of a hero who travels to the glittering gardens of the gods to discover the secret of eternal life. Gilgamesh learns that there *is* no secret: embrace this life, his guide tells him, for nothing else awaits. "Enjoy your life," says the goddess Siduri. "Love the child who holds you by the hand, and give your wife pleasure in your embrace." He returns to the human world, having failed to attain immortality but blessed with an understanding of the beauty of life on earth. Homer's *Odyssey* is said to have been inspired partly by Gilgamesh.

Nearly every image of heaven we hold in our minds—throne room, banquet, and singing angels; garden, rivers, and sweet-smelling flowers; sparkling city, jeweled walls, golden streets; harps, clouds, blinding light; saints and martyrs gathered in worship—comes from such firsthand accounts. Over time, the images have evolved and their emphasis has changed. Travelers to heaven used to see the throne of God, then the saints and martyrs, then characters from history, royal courts, and lush gardens; now, like Piper, they see grandparents and high-school football coaches. In this chapter, I also include artists (writers, painters, movie directors) among the visionaries, since imaginative renderings of heaven—based, however loosely, on the firsthand experiential accounts—have influenced our conceptions most of all.

MEDIEVAL VISIONS

In the earliest apocalyptic literature, as we've seen, visionaries went to heaven, looked around, and received from God a lesson about the fate of the earth. In these visions, heaven is the terrifying throne of God, the seat of Judgment. It is neither pretty nor comforting. The first book of Enoch, written by multiple authors over hundreds of years ending in the second century CE, is a Jewish apocalypse—a glimpse of things to come for those who obeyed—and defied—the law of the Torah. The narrator—named Enoch after the character in Genesis who "was no more, because God took him"—recounts his trip to heaven the way a young boy would tell a ghost story. First, there's mist. Then lightning. Then, "I proceeded until I came near to a wall, which was built of hailstones, and a tongue of fire surrounded it, and it began to make me afraid." Enoch, our hero, faints. The lesson of Enoch, said the scholar James Charlesworth at Princeton Theological Seminary in a phone call, is clear: "God lives in another region, called heaven, and Enoch alone is allowed to go there. . . . He reports that those who are righteous—and suffering—will be rewarded."

John, the author of Revelation, is guided to heaven by an angel and upon arrival immediately sees the Throne. "Around the throne are twenty-four thrones, and seated on the thrones are twenty-four elders, dressed in white robes, with golden crowns on their heads. Coming from the throne are flashes of lightning, and rumblings and peals of thunder." John sees a great battle with Satan, the destruction of the earth, and finally "a new heaven and a new earth." In the last paragraphs, he has a talk with Jesus, who claims to be the one who can turn everything upside down (or right-side up, depending on your point of view). "I am the Alpha and Omega, the first and the last, the beginning and the end." The end is coming soon. John ends his story by swearing its truth. Let anyone who embellishes or erases any part of this tale be among those sent to hell, he says.

The apostle Paul deflected questions about what future was in store by saying it was too much for the puny human mind to imagine: "No eye has seen, no ear has heard, no mind conceived what God has prepared for those who love him." But Paul knew what heaven looked like: according to the Second Letter to the Corinthians, he had been there. Fourteen years ago, he went to "the third heaven," he says, "whether in the body or out of the body I do not know; God knows." What was it like? Paul said only that he "heard things that are not to be told, that no mortal is permitted to repeat." For centuries, Christian ministers have echoed Paul when they tell their flocks that words cannot express the wonders of heaven.

But humans can be literal-minded, and the gaps in the apostle's narrative proved too tantalizing. At least one storyteller tried to fill them in. *The Apocalypse of Paul* was a fourth-century best seller, translated into most European languages. Written by an unknown author, posing as Paul, it tells of Paul's trip to heaven.

Guided by an angel, Paul first watches the souls of the good and those of the wicked leave their bodies. Good souls are greeted upon death with a kiss from a guardian angel and a reminder to remember their bodies, for they will need them at the end of time. Together, soul and angel ascend to the throne of God to await the soul's judgment. The angel vouches for the soul, and then God hands down his ruling. "In as much as this man did not grieve me, neither will I grieve him; as he had pity, I also will have pity." Angels, archangels, and cherubim sing.

In the third heaven Paul sees a great golden gate, and "columns of gold . . . full of golden letters." Upon each of the letters is written the name of one of the righteous; the angel tells Paul that the fate even of the living is recorded in heaven. In the third heaven, Paul sees an old man, whose "face shone like the sun"; they embrace and Paul is introduced to (none other than) Enoch, "the scribe of righteousness."

The next level is a kind of holding pen for the righteous, a heaven before the end of time. Called "the land of promise," it has all the attributes of what many people today imagine heaven to be. On the banks of a river flowing with milk and honey grow trees that bear fruit every month; the whole place shines with a light "seven times brighter than silver." Souls reside here without their bodies. Paul's guide then escorts him to a golden ship, which sails, in the company of three thousand angels, to the City of Christ, the place where the saints, martyrs, and patriarchs wait for the Second Coming of Jesus. Here are the twelve gates and four rivers of Revelation, and here is the great King David, psalter and harp in hand, singing hallelujah and waiting for the return of the Christian Lord. In this version of events, the Jews are singing praises to Jesus in heaven.

From the earliest decades of Islam, Muslim poets and storytellers have also told of a trip to heaven. These are not visions, exactly, but retellings of Muhammad's "Night Journey," an out-of-time trip from Mecca to the place that is now the Dome of the Rock in Jerusalem— on the same hill where the Jewish Temple allegedly once stood. Based on a snippet of the Qur'an—"Glory be to Him Who carried His servant by night from the Sacred Mosque to the Furthest Mosque, whose precincts we have blessed, to show him of Our wonders!"—Muslims have traditionally understood that from Jerusalem, Muhammad ascended through seven levels of heaven until he finally met with God.

The story, given in the hadith tradition, goes like this: Muhammad is sleeping in a cave when he is awakened by an angel named Gabriel who has with him a white, winged steed, bigger than a donkey but smaller than a mule. They fly through the night until they arrive in Jerusalem, where they ascend to heaven on a staircase with alternating gold and silver steps. In the first heaven, Muhammad sees Adam. In the second heaven he sees Jesus and John the Baptist; in the third, Joseph. In the fourth, fifth, and sixth, Muhammad sees—naturally—Enoch, and then Aaron and Moses. Finally, in the seventh heaven, Muhammad sees Abraham, whom Muslims also revere as a patriarch; a big tree covered with butterflies; and a heavenly house, where every day, seventy thousand angels perform something like the hajj—they enter the house, then leave, never to return until the day of resurrection.

Beyond heaven, Muhammad enters paradise, populated mostly with the poor. In some tellings he sees houris there, and celestial servants—ten thousand per resident—holding gold and silver trays. Finally, Muhammad sees the Throne of God. It is supernaturally large, so much bigger than the heavens themselves that it both covers and encompasses them, as the oceans do a mustard seed. It is not for sitting. It is a symbol of God's power.

During the Middle Ages, as such Christian theologians as Thomas Aquinas sorted the sinners and the righteous into their various categories, the heavens of Christian visionaries became more stratified. The three heavens of Paul become a seemingly endless succession of walled-off sections and subdivisions in "The Vision of Tundale," a twelfth-century legend. Tundale was so popular that for centuries it was considered pleasure reading and not just an edifying religious tract, says Eileen Gardiner, editor of a collection called *Visions of Heaven and Hell Before Dante*. In Tundale, the narrator is a knight who has had a stroke and is guided by an angel through layer upon layer of heaven (and hell). As Tundale ascends, the righteousness of the creatures he encounters increases.

After a visit to hell, in which Tundale encounters Lucifer himself, a monster with a thousand hands, he and his angel guide begin to climb. First, they encounter a place for the "not very evil." The stench of hell is gone, and the inhabitants—though hungry, thirsty, and sad—are not in torment. Then they pass through a place for the "not very good," where they see "a beautiful field, fragrant, planted with flowers, bright and very pleasant, in which there was a multitude of souls that no one was able to count." Further on, the place for the faithful married had such a "sweet and delightful odor" that Tundale begged to stay, but his guide urged him forward toward the superior heavens for martyrs, virgins, monks, and the builders of churches.

In the heaven of the monks and the virtuous, Tundale sees one of my very favorite celestial images: a community of tents and pavilions, all in gray and purple silk. From inside the tents, Tundale can hear the strains of delicate music: strings and organs, drums and zithers. Above him, hanging from the sky, are gold chains from which dangle goblets, sweet-smelling flowers, bells, and golden globes. When the angels fly among these celestial wind chimes, "they produced the softest and sweetest song."

Although Tundale sees heaven layer by layer, at the end of the story he also sees it whole. As if from a spaceship, he can see armies of virgins and martyrs, all clothed in white and singing praises to the Lord. The angel guides Tundale back to his body, where he wakes from his coma. He gives all he has to the poor and enters a monastery.

Gardiner, a scholar who has been studying medieval visions for two decades, runs her own small publishing company in Manhattan. She reminds me that these medieval visions—and even the early Jewish and Christian visions—were written, like Piper's books, as *true stories*. These are not artful metaphor; there is no "as if" here. These are powerful, literal accounts of events that the tellers say really occurred—not unlike accounts by people who have seen UFOs—used by believers to underscore the reality of worlds beyond this one. Don Piper is exactly this type of visionary. He saw what he saw, he believed it to be real, and he returned to earth changed, an evangelist determined to tell his story to the world.

PARADISO

Dante Alighieri changed the rules of the game. His *Divine Comedy* is both things—a true vision in the tradition of its precedents *and* a work of art. The pilgrim on the journey through hell, purgatory, and heaven is Dante himself. He writes his story in the first person; he uses his own name. Other medieval visionaries arrive in heaven after a near-death experience—a stroke (as in Tundale) or an injury. Dante's vision is triggered by depression, a sense of being lost, from which he is rescued by Virgil, the Latin poet and his guide through hell and purgatory. He asserts his account's veracity throughout, with claims like "It happened; I swear to you"; certainly some of his readers thought—for real—that the poet had been to hell and to heaven. His near contemporary Giovanni Boccaccio tells this story of a group of Veronese women who spied Dante in the street: "Do you see the man who goes down into hell and returns when he pleases, and brings back tidings of them that are below?" one asked. "To which one of the others naively answered, 'You must indeed say true. Do you not see how his beard is crisped, and his color darkened by the heat and smoke down there?' "

Sensational immediately upon publication (*Inferno* was in wide circulation by 1315, *Purgatorio* by 1320; *Paradiso* was finished the year Dante died in 1321), the 14,233-line *Divine Comedy* was transfiguring. Unlike Tundale, which was intended to be news, Dante's poem was news and art, and as such, it opened the way for the generations of imaginative accounts of heaven that followed. Many readers believe Dante's most astonishing innovation is his choice of heavenly guide, for she is no scriptural angel. Instead, the person who leads the poet through the nine spheres of heaven is a young woman named Beatrice—named after a girl whom Dante allegedly glimpsed on the street in Florence. They exchanged smiles, the story goes, but she married someone else. (So did he.) Beatrice died young, and Dante's love went unfulfilled: "Nothing else in Western literature," writes the literary critic Harold Bloom in The Western Canon, "in the long span from the Yahwist and Homer through Joyce and Beckett, is as sublimely outrageous as Dante's exaltation of Beatrice. . . . Beatrice is the signature of Dante's originality, and her triumphant placement well within the Christian machinery of salvation is her poet's most audacious act."

Paradiso has influenced our images of heaven almost as much as Revelation—though in modern times, it is more frequently assigned to university students than taken up for pleasure by mainstream readers. Paradiso affects our visions of heaven because so many of the West's greatest poets, painters, and writers thought it inspiring and magnificent, among the greatest achievements of Western art. It is difficult going, full of veiled references to literature, history, and Scripture—a knotty theological conversation about the nature of God in terza rima (three-line rhymes with an internal rhyming structure). My own volume has nearly three times as many pages of notes as it does of poetry; it's no wonder Paradiso lives on mostly in graduate seminars.

Michelangelo adored Dante and wrote sonnets to him: Dante "rose a living man to gaze on God." It was, perhaps, the poet's own years-long effort to create with words a picture of heaven and hell—massive, stratified, both human and divine, and populated with characters both famous and unknown—that inspired Michelangelo two hundred years later to do something similar on the walls and ceilings of the Sistine Chapel in Rome. At around the same time, another Florentine, Sandro Botticelli (known better for his erotic paintings of pagan goddesses) made the connection between Dante's vision and early Renaissance art more explicit. He created nearly a hundred drawings of *The Divine Comedy*, exquisite in their simplicity, for the great arts patron Lorenzo de' Medici. In *Ascent to Primum Mobile*, which looks like a modern sketch, Dante covers his eyes as his guide Beatrice points skyward toward the God-light, a sun surrounded by nine circles of stars. In 1667, John Milton published his own heavenly epic, of course: *Paradise Lost*. The British poet and illustrator William Blake spent the last years of his life, in the 1820s, illustrating *The Divine Comedy*. Henry Wadsworth Longfellow, James Joyce, the mystery writer Dorothy L. Sayers, the playwright Samuel Beckett, and the contemporary poet Seamus Heaney—all have loved Dante and credited him as inspiration. (Beckett once told a friend that all he wanted to do was "sit on my ass, fart and think of Dante.") In a poem called "Station Island"—named after the location of St. Patrick's Purgatory mentioned in the previous chapter—Heaney uses *terza rima* to evoke the bright-light vision of *Paradiso* in modern, murkier terms:

> As if the prisms of the kaleidoscope
> I plunged once in a butt of muddied water
> Surfaced like a marvelous lightship

"How many ways can you say 'light'?" complains Jean Hollander, as I sit in her Princeton, New Jersey, kitchen with her husband, Robert, and their rambunctious German shepherd, Josie. The Hollanders had, in the weeks before my visit, just published their joint translation of *Paradiso*. Robert, the Princeton scholar, knew the poem as well as he knows his own name—its sources, its history, its scholarly controversies and, of course, the medieval Italian dialect in which it was written, the language that became modern Italian. Jean is a contemporary poet. Together they created a sophisticated poetry translation—with ample notes—that modern readers can still find beautiful. One of my favorite passages is this, in which Dante asks Beatrice whether a person might stint on certain good works and substitute others and still gain salvation:

> Beatrice looked at me with eyes so full
> Of the radiance of love and so divine
> That, overcome, my power of sight faded and fled,
> And, eyes cast down, I almost lost my senses.

Beautifully done, but the problem of all that light obviously rankled Jean. "All anyone ever does is stand around and look at the light." She reveres Dante, but his vision of heaven doesn't sound like a place to which she'd want to go.

Dante's heaven is a place of light, sweet smells, and music. At the beginning of *Paradiso*, light is like air, part of the atmosphere. "It seemed to me that we were in a cloud, / shining, dense, solid, and unmarred, / like a diamond struck by sunlight. / The eternal pearl received us in itself / as water does a ray of light / and yet remains unsundered and serene." As Dante ascends through the spheres, guided by Beatrice, he meets saints and martyrs, people he knew in life, and people he knows from history. The light grows brighter and brighter, and in the third-to-last canto, when he reaches the Emyprean heaven, Beatrice bows out of her role as his guide and sits down. Dante addresses the light directly. Atmospheric light, with a lowercase *l*, becomes capitalized—the "Light" of God. Dante looks into the Light and says he has finally arrived. "I reached the Goodness that is infinite."

Inside the Light, Dante sees that all the variety in the universe is bound together by love, and he understands the wholeness of everything. The Light then changes color and becomes three circles, the Trinity, and as Dante gazes upon it he sees that it is "painted with our likeness." It is light, it is color, it is three, it is one, and it looks like us. As he's struggling to understand what he sees, his mind is "struck by a bolt / of lightning that granted what I asked." He grasps something divine, but like Paul, Dante insists that words fail him in describing his vision—that trying to remember it is like trying to remember a dream. That bolt of understanding changes him so he is no longer separate from the universe but one with it—and then he goes home. "My will and my desire," he concludes, "were turning with / the Love that moves the sun and all the other stars."

Robert Hollander would say that Dante saved his life. He is not exaggerating. He and Jean had finished their translations of *Inferno* and *Purgatorio* and were nearly through with *Paradiso* when Robert had a massive stroke. His brain could not retrieve any words in English—let alone in Italian. Several weeks into his recovery, Jean brought to the hospital their translation of *Paradiso*, and Robert began working on it, word by word, line by line. "It was like waltzing in mud," he told me. But the *Commedia* has been Robert's life's work—he fell in love with Dante while a young teaching assistant at Columbia, and has never recovered. *Paradiso* has been his sustenance and inspiration for more than forty years. And although he does not believe in an afterlife, he wishes he did—and if he did, it would look exactly as Dante describes it in *Paradiso*.

In Canto 23, the Virgin appears to Dante, and she is surrounded by whirling music and flames. Her head is encircled by a sapphire crown. Dante reaches toward her "like a baby reaching out its arm / to *mamma* after it has drunk her milk." The saints, dressed in white, reach toward her, too, singing "*Regina celi* with such sweetness / that my feeling of delight has never left me." Unseen, but presiding over all, is Jesus. Hollander's body sags in his chair, but his dark eyes are alight with intelligence and love. "I think he's got it right. It's absolutely convincing to me. If I'm wrong and nonetheless saved, that's what I'll see. It's a tempting thing. If I ever convert, it will be because of *Paradiso* 23."

ARTISTIC VISIONS

With his example, Dante gave generations of artists permission to imagine heaven as they wished—and in the decades after the celebrity-poet's death, Italian painters envisioned it just as Dante did. Frescoes of the Last Judgment in the Strozzi Chapel in the church of Santa Maria Novella in Florence depict the heaven—and hell—of the *Commedia*. Saints are seated in happy rows; Dante himself is among the elect. In the early 1300s, the Paduan aristocrat Enrico Scrovegni commissioned the painter Giotto to render the Last Judgment on the walls of the Arena Chapel, and Giotto used as his template the *Commedia*. The saints, again arrayed in tiers, all gaze at Christ, encircled by light. Not by chance, Enrico himself kneels before the Virgin Mary at the bottom of the picture, offering his chapel as a gift to her: Dante had consigned Enrico's father, Reginaldo, to the seventh circle of hell for usury. By sponsoring the picture, the son was endeavoring to redeem his father's sins—and reputation.

But as the Renaissance bloomed, painters began to feel freer with their imaginative visions. I met Ena Heller in her office at the Museum of Biblical Art (MOBIA) in Manhattan. The executive director there,

she specializes in early Renaissance art, and over coffee one morning
we gazed at one of her favorite depictions of paradise—which she pulled
up on her desktop computer—created by an unknown Rhineland
painter around 1440. In it, a low stone wall encircles a blossoming
garden of fruit trees, iris, and lilies. Birds flit in and out of shrubs. The
Virgin and saints are seated or standing in the grass. A baby plays a
lyre. In the Renaissance, "the heavenly narrative becomes more detailed and complex," explains Heller, "and they go much less by the
strict standards mandated by the Middle Ages." She pulls up another
favorite image on her screen. This one is a sixteenth-century fresco on
the exterior wall of a church in Moldova. It shows an old man, with
white hair and a beard—this is Abraham—rocking a handful of min-
iature white-clad souls, like babies, to his breast.

With the Reformation, Heller explains, images of heaven—and
saints and angels—came down from the church walls. Calvin, especially, discouraged every kind of imaginative rendering; paintings,
frescoes, tile work, stained glass were seen as corrupting influences.
"Your connection is to be with the word, the word over images. You
just need to read the Bible, and you shouldn't imagine it," Heller
explains. Heller—who was raised in the Eastern Orthodox tradition—
grew up in churches covered from floor to ceiling with gold icons.
The first time she stepped into a church in Switzerland, she says, "I
stood there thinking, 'How can people pray?' I can't imagine ritual
without images—just not to have it on the walls, it was really weird
to me."

This very human rebellion against the Reformation—with all its
abstractions and austerity—produced the Counter-Reformation, and
with it a renewed appreciation in the Roman Catholic Church of music,
culture, and art. Many of the images we have come to associate with
heaven today come from that period. In the *Assumption of the Virgin*

Mary, Peter Paul Rubens's seventeenth-century Antwerp altarpiece, pink, fat-faced angels with feathered wings whirl and whoosh around the Virgin as she, smiling, floats upward, her skirts moving with her. Beams of beckoning light gleam from above. This is the heaven of Christmas cards and of our conventional contemporary imaginings. Popular visions—in painting, at least—ceased evolving about a hundred years later during the Enlightenment just as the Church stopped patronizing the arts, Heller explains. "There's nothing really new that gets created. That's why these images have such enduring power. There wasn't anything to replace them."

Today, artistic visions tend to be idiosyncratic. In the modernist German artist Anselm Kiefer's 1990 book *The Heavenly Places: Merkaba* (the name of a school of early Jewish mysticism in which visionaries ascend to heaven and return to earth with eyewitness accounts), he presents page after page of black-and-white photographs. First he shows a bare, cellarlike space strewn with rubble and dotted with pillars. As the book continues, the images gradually resolve into a woman screaming (either in ecstasy or horror). *Book with Wings* is a sculpture: a volume of Scripture looks ready to fly, thanks to the enormous angel wings attached to each side. In all religious traditions, of course, Scripture is itself a door to heaven—and for mystics, chanting, praying, reading with steady repetition, one gets closer to God. Kiefer, who was raised Roman Catholic, turned in later life to Kabbalah and other forms of mysticism. "Heaven," he told the *Times* of London in 2007, "is in each of the little sparks of the body."

Writers, too, continue to try their hand at heaven. In 1946, C. S. Lewis wrote a visionary journey to heaven called *The Great Divorce*, and in it articulated the interpretive problem that Dante left implicit. For the believers, even the highest literature leaves much of heaven unimaginable: "Do not ask of a vision in a dream more than a vision in a dream can give." Alice Sebold and Mitch Albom followed—but the medievalist Gardiner suggests that some of the most interesting recent results have been in the realm of science fiction. Ursula Le Guin's short story "Paradises Lost" from her collection *Birthday of the World* has a plot rather like the 2009 Academy Award–winning animated movie *WALL-E*: aboard a spaceship on a two-hundred-year journey, the passengers begin to believe they're in heaven. None of earth's problems affect them anymore. There's no hunger—the spaceship crew instantly gives them everything they need. There's no disease, no war. When they arrive at their destination, the passengers must decide whether to disembark onto the "dirtball" on which they've landed—or to stay aboard. "Did our ancestors send us from one hell to another hell by way of heaven?" asks one.

If you want to create an enduring and salient image of heaven for the masses these days, you'd best make a movie. With *Defending Your Life*, the director Albert Brooks in 1991 produced and directed one of the twentieth century's most inventive afterlife visions—although *What Dreams May Come*, from 1998, gets an honorable mention for special effects. The hero, Daniel Miller (Brooks), dies unexpectedly and ascends to a kind of celestial way station—not heaven, exactly—where he has to defend his life in a courtroom setting. The judgment of the court determines his final destination. Miller has led an unexceptional life, marked by conservatism, fear, and anxiety—and in retrospect, seen in short recorded flashbacks, it all looks bad. During his stay, he meets Julia (Meryl Streep), whose life has been brave and glorious: she was the kind of woman who rescues children and pets from burning buildings. They fall in love, and through love, Miller grows brave—and as the Mormons would say, "worthy" of ascending to the next level.

Like *Paradiso*, *Defending Your Life* is a complete imaginary vision of the afterlife. Brooks created a whole world, in which electric trams shuttle the dead back and forth between their court appointments and the hotels where they temporarily reside; in which the food you order at restaurants arrives at the table before you're finished ordering and is always delicious; in which every resident wears a white garment called a tupa, a cross between a hospital gown and a toga. Brooks even envisioned entertainment for the dead—floor shows, stand-up comedy, or, in the Past Lives Pavilion, a retrospective look at your previous incarnations.

I spoke to Brooks one evening by phone. Nearly twenty years on, he is still extremely proud of *Defending Your Life*, and Daniel's late arrival at self-knowledge and emotional bravery can make him cry when he catches the film unexpectedly on television. Brooks says he knew *Defending Your Life* would succeed only if he established every detail of Judgment City: what everything looked like and how things worked. "It took a lot of thought over a long period of time," he says. "I had these charts all over the place. I had to make up all these rules. How long you were there, for instance. You could eat all you want. . . . How does that work? How much arguing could you do in the courtroom? It had to be made up from scratch—the transportation system, what you would feel, what you wouldn't feel. . . . If there was consciousness, what consciousness? The tupas—the idea of what you wear and what that looks like. You can't hire a costume director and say, 'Do what you did in *Titanic*.'"

Brooks has not read Dante, but he has a similar interest in death and justice—and the possibility of justice after death. He is Jewish. He had a bar mitzvah, goes to temple on the High Holy Days, and lights *Yahrzeit* candles in memory of those he's lost. "The one thing I like about Judaism very much, what I gained from it, was the Golden Rule," he tells me. "That always made sense to me. I don't want to be shitty to people because I don't want them to be shitty to me."

Like many Americans, Brooks is of two minds about heaven. His father died when he was just twelve. "The percentages say nothing happens to you after you die. We can see that any night we fall asleep and don't dream. That could go on for eternity. But, having said that, I don't know. Maybe. It would be great if the movie was right, if someone said, 'You still have a lot of fears, but you figured it out, so come with me.' . . . If I am right, I'd like to be awarded a posthumous Nobel Prize. More than that. Free parking."

NEAR-DEATH EXPERIENCES

Carol Zaleski is the mother of modern heaven studies. A practicing Roman Catholic, she became fascinated with heaven as a young child. "Heaven has always been the best, most perfect place people can imagine," she wrote in an e-mail. "Who wouldn't want to think about that?" Now a religion professor at Smith College, she and her husband, Philip, have together written or edited a number of books relating to heaven, including an anthology called *The Book of Heaven* (which I thumbed heavily during my research) and a book on prayer. Her most intriguing book by far, however, is the one that came out of her doctoral dissertation at Harvard University. *Otherworldly Journeys: Accounts of Near-Death Experience in Medieval and Modern Times*, published in 1986, puts the popular visions of antiquity and the Middle Ages in the same category as near-death experiences (hereafter referred to as NDEs), though she adds that both need to be interpreted with care. Some are touchy about this analysis, for they feel it undermines the truth of their experience. Don Piper explicitly denounces those who say that what he saw was merely an NDE. "I didn't undergo anything like [that]," he writes in *Heaven Is Real*. "One second I was alive, and the next instant, I stood at the gates of heaven." Nevertheless, it's a connection worth exploring.

I met Zaleski one winter afternoon in the beginning stages of researching this book, in a coffee shop in Northampton, Massachusetts. It was the Christmas season, and the streets outside were covered with slush. In the bustle of the shop, Zaleski's calm, placid manner struck me—as if she herself had seen heaven and come back, changed. In a sense, she has. Zaleski grew up in a nonobservant Jewish household in Manhattan, amid tall buildings and other families like hers who "had ham for Easter dinner." Her lack of formal religious training left a hole, a sense of something missing, and so she began to seek to fill it. She read poetry books given to her by her grandmother—William Butler Yeats and Allen Ginsberg. She volunteered downtown at the Catholic Worker Movement, the social justice group founded by Dorothy Day. She befriended a group of Benedictine monks in northern Massachusetts. In 1991, at the age of forty, she was officially received into the Roman Catholic Church. Her conversion was a long, slow process that came about after decades of resistance—she didn't want to betray her roots. She believes in heaven—she likes the idea "that a perfect life is available to us, a more meaningful cosmos than the world allows." Her favorite vision of heaven, she says without hesitating, is Dante's *Paradiso*—"definitely."

In modern America, with all the resuscitation techniques available to us, the number of NDEs is surprisingly high. According to a 2000 article in *Lancet*, between 9 and 18 percent of people who have been demonstrably near death report having had one. Surveys of NDE accounts show great similarities in the details. People who have had NDEs—like some religious visionaries—describe a tunnel, a light, a gate, or a door, a sense of being out of the body, meeting people they know or have heard about, finding themselves in the presence of God, and then returning, changed. Such experiences are common across cultures and religions, although they're interpreted differently. Zaleski warns against taking these visions too literally and speaks movingly of what Catholics call "discernment," the process—with divine help—of deciding what is true. "There are a lot of clues to authenticity," she says. "Does the experience shed more light? Does it make you more loving? Does it have a quality of reality to it? Does it connect you with other things you value? If yes, then I'd say go with it, but wear it lightly even so."

Skeptics might frame this question of authenticity differently: Can these visions be explained simply through the chemical and physiological processes of dying? Or have people who have had NDEs really made some kind of extra-conscious journey? (Or, to get really tricky, is this a meaningful distinction?) This is a difficult field of study for obvious reasons: most people in the process of dying *do* die, and those who survive approached the brink of death in such different ways—car accident, stroke, heart attack—that it's impossible to compare their experiences scientifically. But over the years, science has posited a number of theories about the connection between visions of heaven and the chemical and physical processes that occur at death.

Andrew Newberg is an associate professor in the Department of Radiology at the University of Pennsylvania and has made his reputation studying the brain scans of religious people (nuns and monks) who have ecstatic experiences as they meditate. He believes the "tunnel" and "light" phenomena can be explained easily. As your eyesight fades, you lose the peripheral areas first, he hypothesizes. "That's why you'd have a tunnel sensation." If you see a bright light, that could be the central part of the visual system shutting down last.

Newberg puts forward the following scenario, which, he emphasizes, is guesswork. When people die, two parts of the brain, which usually work in opposition to each other, act cooperatively. The sympathetic nervous system—a web of nerves and neurons running through the spinal cord and spread to virtually every organ in the body—is responsible for arousal or excitement. It gets you ready for action. The parasympathetic system—with which the sympathetic system is entwined—calms you down and rejuvenates you. In life, the turning on of one system prompts the shutting down of the other. The sympathetic nervous system kicks in when a car cuts you off on the highway; the parasympathetic system is in charge as you're falling asleep. But in the brains of people having mystical experiences—and, perhaps, in death—both systems are fully "on," giving a person a sensation both of slowing down, being "out of body," and of seeing things vividly, including memories of important people and past events. Does Newberg believe, then, that visions of heaven are merely chemical-neurological events? He laughs nervously. "I don't know." He laughs again. "It's, um. . . . I don't think we have enough evidence one way or another to say."

Since at least the 1980s, scientists have theorized that NDEs occur as a kind of physiological self-defense mechanism. In order to guard against damage during trauma, the brain releases protective chemicals that also happen to trigger intense hallucinations. This theory gained traction after scientists realized that virtually all the features of an NDE—a sense of moving through a tunnel, an "out of body" feeling, spiritual awe, visual hallucinations, and intense memories—can be reproduced with a stiff dose of ketamine, a horse tranquilizer frequently used as a party drug. In 2000 a psychiatrist named Karl Jansen wrote a book called *Ketamine: Dreams and Realities*, in which he interviewed a number of recreational users. One of them, who called himself K.U., describes one of his drug trips this way: "I came out into a golden Light. I rose into the Light and found myself having an unspoken interchange with the Light, which I believed to be God. . . . I didn't believe in God, which made the experience even more startling. Afterwards, I walked around the house for hours saying 'Mine eyes have seen the glory of the coming of the Lord.' "

To some people, the explanation of heaven as an event that happens in your brain as you're dying is enough. My friend Christopher Dickey is one of these people. He told me that his father, the writer James Dickey, had a recurring dream in which heaven was like a swimming pool around which his friends sat, chatting. "There was nothing special about the pool itself," wrote Chris in *Summer of Deliverance: A Memoir of Father and Son.* "Nobody walked on the water. And he never told me who the friends were. . . . But what he took away from the dream was a sense of contentment, of being at ease with himself and the world, as if he had gotten a preview of heaven. He called that place "The Happy Pool." Chris recounted this anecdote to me with a sweet smile and then said he imagined that everything we think we know about heaven happens in the moments before death. After that, there's nothing. (The narrator of Philip Roth's *Indignation* is dead—or rather, thinks he's dead—and he believes much the same thing. Death is not "an endless nothing," says Roth's Private Messner, but "memory cogitating for eons on itself.")

Others, though, remain dissatisfied. Emily Williams Kelly is a psychologist who works at the University of Virginia's Division of Perceptual Studies, which treats the study of NDEs as a legitimate science. Her résumé is impressive: she has degrees from Duke, UVA itself, and the University of Edinburgh—not institutions one usually associates with the study of the supernatural or paranormal. Kelly has spent her career researching, as she puts it, "the interface between the brain and the mind." Practically speaking, she interviews dying people and tries to find patterns and similarities among their experiences. Kelly believes the experience of people who have had near-death visions demonstrates that consciousness exists even after normal brain function ceases— a theory that could suggest explanations for an afterlife beyond the scientific. "If our conscious experience totally depends on the brain, then there can't be an afterlife—when the brain's gone, the mind's gone," she says. "But it's not that simple. Even when the brain seems to be virtually disabled, people are still having these experiences."

What is she saying? That upon death people really go to another realm—and that science can prove it? Kelly shrugs. NDEs, she says, "tell us to open our minds and think there may be a great deal more to mind and consciousness—that's as far as I'm willing to go. Whether it means they've actually been to heaven—I couldn't go that far. . . . But our current model of how the brain works doesn't explain it." The frequency and uniformity of NDEs "opens up the possibility that these experiences are what they say they are, and we have to take that seriously."

Mally Cox-Chapman would agree. We are sitting at her kitchen table, in her large, comfortable house in Hartford, Connecticut, and she is using an analogy to explain the way our limited minds think about heaven. One of her hands is in a fist on the table. That's "space bug." The other is flat, palm down. That's "flat bug." Together, they cruise, side by side, toward an obstacle—a coffee mug—before them. Flat bug, who's two-dimensional, can't imagine how to get around the obstacle. He only thinks in two dimensions. But space bug just hops across. Easy. He exists in space. Why is it so hard, Cox-Chapman wonders aloud, to imagine a soul discovering another dimension? Cox-Chapman is a believing Christian, and she wants to convince me that a supernatural dimension created by God is, at least, a possibility.

Cox-Chapman is the author of *The Case for Heaven*, a collection of interviews with near-death experiencers published in 1995. Her argument: that experiences of the millions of people who have been to the other side corroborate the claims of faith. According to their own testimony, they come back changed. Unbelievers find belief; travelers say the memory of their NDE never fades; some learn things in the afterlife they didn't know before. Cox-Chapman believes there must be a portal—or a gate, a fence—and through it must be heaven, a place, as the accounts testify, of love, light, and overwhelming joy. Audience ratings soared when she appeared on *Dateline NBC* in 1996.

She herself had what she calls "close to a near-death experience" in 1983, while she was driving in the dark and the rain near the Schuylkill River in Philadelphia. Her car hydroplaned and started to spin. As she careened toward the river, Cox-Chapman heard a voice coming from somewhere near the sun visor. "Relax," it said. "It's not your time to die." Cox-Chapman relaxed and let go of the wheel. (I am reminded here of the 2005 Carrie Underwood hit single, "Jesus, Take the Wheel," a country song with a similar plot line.) Within seconds, her car had stopped, up to its chassis in mud, feet from the river. Cox-Chapman walked away from the scene utterly calm and unbruised. "Some people would call that disassociation," she says. "Some might wonder if there's been an angelic presence."

The memory of the accident and her miraculous escape stayed with her, and a decade later Cox-Chapman began work on a book about near-death experiences, which she collected by the dozens. Anxious not to appear discreditable, Cox-Chapman reminds me, again and again, of her secular-elite credentials. She went to Yale University. She married a successful physician. She worked for years as a freelance journalist. She is not unhinged, she assures me. She simply can't refute what she sees as incontrovertible evidence: heaven must be real. In her first chapter, she tells the story of Mary Dooley, a fashion designer who went to heaven during her surgery for uterine cancer. She leaves her body and sees a bright light ahead; Dooley chides herself at first for having forgotten her sunglasses. "However," she says, "when I reached the brightness, I found no strain to my eyes, nor did I need any adjustment for the light. Though it was extremely bright, there was much softness to it. *Soothing* would be a good word." Later Dooley feels the presence of God. "His radiance was everywhere," she says, echoing Dante. "He *was* the light."

I find myself staring into space in Cox-Chapman's cozy kitchen, at a child's painting that's a decade old. I am confronted, again, with the question of what I believe. I believe the believers, I tell Cox-Chapman, slowly. I believe they saw what they said they saw. But I don't believe that their testimony, as consistent and thrilling as it is, adds up to proof that there's a heaven. I am much more comfortable, I tell her, with the idea of heaven as a matter of faith, not proof. I think of the words of Maimonides, who echoes Saint Paul: "As to the blissful state of the soul in the World to Come, there is no way on earth in which we can comprehend or know it." I can tell that I've disappointed her.

EIGHT

REUNIONS

Elaine Hasleton is a small, neat woman in her mid-fifties who runs the public relations office for the Family History Library, the five-story genealogy research center belonging to the Church of Jesus Christ of Latter-day Saints in Salt Lake City. She, like so many Americans, believes that after she dies she will see beloved family members in heaven. But when Hasleton says "family," she doesn't just mean the people who live in her house or the people she sometimes visits on Thanksgiving. She means twelve generations of relatives, living and dead, spread out across two continents, mostly in Norway and Minnesota. She means Norwegian musicians and house painters and farmers dating back to the 1500s. Since her graduation from Brigham Young University in 1972, she has been to Norway eleven times and has tracked down countless cousins—including a group who share a common ancestor back to 1793. She obsessively reads obituaries and recently found a cousin in Minneapolis who is related to her great-grandmother, Henrietta Krogstad, the wife of a Norwegian tailor. This was especially fortunate because the tailor, Ingebret Krogstad, is the only person whose portrait is missing from Hasleton's illustrated five-generation family tree, and Hasleton thought this new cousin might lead her to one. She didn't, but in any case Hasleton believes she will finally see Ingebret's face in heaven. "I can't wait," she says.

Mormons believe that after death, the righteous will live in multi-generational clans: spouses together with children, cousins, aunts, uncles, grandparents, and great-great-grandparents. Young men who died in war, before they ever had a chance to marry, will be wed in the spirit world—the stop before the resurrection, when everyone goes with their families to heaven. Women who died alone and childless through no fault of their own will have spouses and kin. Everyone will be bound together, as the Mormons themselves put it, in a "chain of generations." The temples and other holy places of the Latter-day Saints are decorated with pictures of families in heaven; the "celestial room" in the Idaho Falls temple shows white-clad men and women lounging or standing in groups of four or five, as though picnicking on golden grass; purple mountains rise behind them. Devout Mormons passionately want this for themselves and their children, and it explains their seemingly obsessive search for ancestors.

But of course to get to the highest level of the Mormon heaven, you have to be a Mormon. You have to believe that the religion's founder, Joseph Smith, was a prophet of God and that the revelations given to him by an angel in upstate New York in the 1820s are true. This means that everyone born before Joseph Smith's time—as well as everyone born afterward who doesn't believe in Smith's prophetic authenticity—needs to be inducted into heaven, as it were, by a living, card-carrying Mormon. It is for this reason that the LDS church in Salt Lake City holds the world's largest collection of genealogical records—birth, death, and marriage records, as well as other documents from 110 countries, dating back hundreds of years. Living Mormons collect the names of dead souls to help them reach the celestial kingdom.

Unearthing names is just the first step. Mormons take the names to the temple and perform such sacred rituals as baptisms and marriages posthumously on behalf of the people attached to those names. In recent decades the Mormons' zeal for ushering souls into heaven has extended not just to their own ancestors but to any soul ever born. In the 1990s, the LDS church came under fire for posthumously baptizing Jews who died in the Holocaust. On a recent visit to Salt Lake City, I initially resisted the urge to search for my own dead relatives: the idea that the Mormons might want to baptize them made me squeamish. Ultimately, though, my curiosity won out. Sitting at a terminal in the Family Research Library, I typed the name of my maternal grandfather, Marcel Goldmuntz, into an LDS computer. My grandfather was born in Belgium in 1911, the eldest son of a prosperous Jewish family, and he died in August 2000, in his bed in New York City.

Punching at the keys, I first found his death certificate. Seconds later I found the passenger manifest of the *Serpa Pinto*, the ship that sailed from Lisbon to New York in March 1941, carrying my grandparents and my three-year-old mother. "Eyes: Brown," the document said by my mother's name, "Race: Hebrew." After a year of driving, traveling, and hiding, they arrived safely in New York.

I have on countless occasions reflected on my grandfather's prescience and bravery, his willingness on that May morning in 1940 to leave everything he had (wedding silver, some beautiful rugs—and, my grandmother always remembered, a cute tennis outfit she had recently bought to celebrate the spring weather). He also left everyone he knew to save the lives of his family. There, in the basement of the LDS church, with two kindhearted Mormon women peering over my shoulder, I lost my composure.

Hasleton assured me later that LDS members would not use my search as an opportunity to seize my grandparents' names and baptize them. (LDS officials add that the rites they perform on behalf of the dead are not binding: souls may still choose whether to enter the celestial realm.) She has enough work to do just gathering her own family together.

Elaine Hasleton has done "temple work" for "just about all" of her dead relatives, "though there's still more to do. It's hard to find the time."

Who do you see in heaven? The Mormons are hardly the only ones who conceive of heaven as a domestic place, like home but better: with no squabbling, no anxiety, no mortgage payments, divorce, or cancer. For many people, heaven is eternal togetherness, a recapturing of the family time that in life can feel so quickly and easily lost. In this vision, you can feel your child's silky hair again; you can hear your husband laugh. The yearning to reunite with loved ones is primal and overpowering: Nearly half of Americans believe they'll see family and friends in the next life, according to a 1982 Gallup poll. It is the thing we tell our children. It is boilerplate in Christian, Muslim, and Jewish funeral sermons. The late Senator Edward Kennedy, a devout Roman Catholic who had too much experience giving eulogies for his relatives, reflected his belief in a heaven of family happiness when in 1999 he spoke at his nephew John F. Kennedy Jr.'s funeral. "He and his bride have gone to be with his mother and father, where there will never be an end to love."

Probably no further proof of the strength of this belief in contemporary America is needed, but here it is: Even Homer Simpson abandons his doofus persona for a split second in episode sixteen of the cartoon series *The Simpsons* to tell God that "heaven isn't heaven without my family in it." In *Heaven*, a book of essays edited by an Episcopal priest named Roger Ferlo and published in 2007, ten notable and unknown Christian writers describe their visions of the hereafter. The descriptions vary, but all convey a longing to see loved ones—and in some cases personal heroes—in the afterlife. "I want to lay my head on Grandma Lucy's lap again," wrote the Christian memoirist Barbara Brown Taylor. "I want to shell field peas with Fannie Belle and listen to Schubert with Earl."

The tough-talking president Lyndon Johnson worried about seeing his parents again in heaven. In an anecdote in *The Preacher and the Presidents*, a biography of the evangelist Billy Graham by Michael Duffy and Nancy Gibbs, Johnson confesses his anxiety to the world's most famous preacher. The president and Graham are driving around Johnson's Texas ranch, and Johnson parks his car by the spot where his mother and father are buried beneath some shady oaks. He then turns to Graham and, after a pause filled with meaning, asks, "Billy, will I ever see my mother and father again?"

"Well, Mr. President," Graham told him, "if you're Christian and they were Christians then someday you'll have a great homecoming."

For believers, the certainty of ultimate reunions offers great solace. Some years ago, I visited Barbara and Warren Perry at their home in Kinston, North Carolina, not long after Warren had received a diagnosis of terminal pancreatic cancer. The Perrys are evangelical Christian believers and friends of Anne Graham Lotz, Billy Graham's daughter. While I was in North Carolina visiting Lotz, she suggested I call them; heaven, she said, would be much on the Perrys' minds. I phoned from the road, and Barbara invited me right over. Sitting on their sun porch, while eating a lunch of turkey soup and melted cheese sandwiches, I asked them all—Warren, Barbara, their daughter Betty Blaine and granddaughter Whitley—what they thought about heaven, and they all began to cry. They would see Warren soon, they said, in heaven. After lunch, Warren went upstairs for a nap, and teenaged Whitley—a lovely blond girl with kohl lining her eyes—told me she was eager to go to heaven herself because, along with family members, she knew she'd see Saint Paul. "That will be so cool," she said.

Warren died six months later, and Barbara vowed she would not remarry. "I know I will see Warren again," Barbara told me later in a phone call. "I had such a wonderful, wonderful marriage, I don't want to mar my reliance on that."

So essential is the family reunion to modern conceptions of heaven that the reader may be surprised to learn that it is a recent development, widespread only in the last two centuries. In antiquity, people may have believed that they would inhabit the same heaven as their loved ones—but they emphatically did not believe that life in the next world would be anything like this one. Intimate, familiar relationships would cease to exist. "There will be no marriage in heaven," Jesus said.

As we saw in the previous chapter, heaven has always been full of people. Christian visionaries saw saints, martyrs, and the biblical patriarchs as they ascended through the levels of the afterworld. The Virgin Mary inhabited heaven, of course, as did her son, Jesus, and the righteous expected to see them there. As if visualizing herself from the outside, the thirteenth-century German mystic Mechthild of Magdeburg described her meeting with Christ in heaven: "He took her into his divine arms, placed his paternal hands unto her breast and beheld her face. And in a kiss she was elevated above all the angelic choirs." According to Muslim tradition, Muhammad on his Night Journey saw Moses, Enoch, and even Jesus as he traveled upward through the spheres.

Antique and medieval believers did imagine their own relatives in heaven. In the tenth century, Saadiah Gaon, a Jewish philosopher living in Baghdad, affirmed his belief that Jews would most certainly meet in the next world. "Should someone ask, again, whether members of their families and their kinsmen would recognize the resurrected and whether they would recognize each other, I might say that I have considered this matter and have arrived at an affirmative conclusion." Saint Augustine in the *Confessions* begs readers to pray for his parents in heaven and, after he becomes Bishop of Hippo, he consoles a grieving widow in a letter with the promise of a new life.

But, Augustine insists, people in heaven will not be focused on one another. They will be focused on God. Happy reunions and the respite of individual loneliness and grief are precisely not the point. "After death shall come the true life," Augustine wrote to the widow, "and after desolation the true consolation." Heaven—in the ancient Christian perspective, at least—is the place (or time) for all believers to collectively and eternally rejoice in the divine glory. Thomas Aquinas wrote that even if heaven were occupied by one single soul, "it would be happy, though having no neighbor to love."

Dante's *Paradiso* has long been seen as the prime example of what scholars call "theocentricism." While peopled with saints and martyrs, with individuals the poet knew as a boy (including a great-great-grandfather), and with historical figures, human relationships are irrelevant. In Canto 32 of the *Paradiso*, Dante describes heaven as a sort of crowded stadium. Souls are arranged in rows; the holiest souls get the best seats, closest to the light, while the least worthy sit farthest away. No matter their rank, all are perfectly content—for all are engaged in the activity of heaven, which is gazing upon God.

Peter Hawkins, who is a professor of religion and literature at the Yale Divinity School, argues, however, that in this scene Dante opens the door a tiny crack to the kind of intimate heaven we imagine today. I met Peter at his Boston condominium one late afternoon and we sat together drinking wine as the sun set over his balcony and the Public Garden nearby. Everyone in Canto 32 of *Paradiso* is looking at God, Hawkins agrees, with one exception. He points me to this verse:

> Look at Anna, where she sits across from Peter
> So content merely to gaze upon her daughter
> She does not move her eyes as she sings hosanna.

Anna, Hawkins explains with excitement, is not looking at God at all. The mother of Mary can't keep her eyes off her own beloved daughter. "She's looking at her girl chick!" Hawkins exclaims. "She's looking at her kid!" At least one person in Dante's heaven puts a personal relationship above the adoration of God. It surprises me not at all that this person is a mother.

After Dante, depictions of heaven became more like human society: hierarchical, to be sure, but also full of all kinds of people—nobles, clerics, merchants, and professionals all intermingled. In paintings, the Virgin Mary—once a formal, queenly presence in heaven—became during the Renaissance more tender and human. In deference to their classical forebears, Renaissance painters endeavored to make people, even people in heaven, look like themselves, and *real*. In Raphael's 1505 *Madonna of the Meadows*, the Virgin looks like a fair-haired and earthy Italian matron; she wears simple garb as she tends to two naked babes—Jesus and John the Baptist—playing in a golden-hued field. Mist rises from a lake in the background; a church spire penetrates an azure sky. In other imaginative renderings, such as Giovanni di Paolo's *Paradise*, painted around 1445, saints, angels, and regular people meet and mingle in a flowering garden. "Irrespective of spiritual merit and state, the blessed . . . mix freely in the relaxed atmosphere of a garden," Colleen McDannell and Bernhard Lang write in their book *Heaven: A History*. "The artists depicted a single paradise where the ordinary person . . . might live." In these pictures, heaven mirrors human society—not yet family reunions or domestic bliss—but a vision of the world to come that resembles life on earth.

But with the Protestant Reformation in the sixteenth century, heavenly society withdrew once more. The reformers "purified heaven," explains Alister McGrath, a theologian at King's College London, whom

I reached one afternoon by phone. In the Reformers' heaven, "there is nothing except the believers contemplating God," says McGrath.

In Puritan New England, a heaven that looked like a society ball or even a church dinner would have been unimaginable. The social life of each town did revolve around its church, but the message people heard on Sundays was dire. Life was a matter of endless toil and continual struggle against the forces of sin and temptation. Heaven was something to look forward to, but it was more abstract—utterly unlike earth and far off in the future. It was not primarily a comfort but rather a way to impose discipline in this life. David D. Hall, a historian of American religion at Harvard, described the Puritan mind-set to me this way. "You're trudging along, trudging along, trudging along. You look upwards, and you're aware of great rewards that are ahead, but they are remote." In a sermon preached at Cambridge, Massachusetts, and published in 1677, the Puritan minister Jonathan Mitchell painted this grim picture to great effect: heaven has "none of that Confusion, Disorder or Discomfort that results from sin and temptation, and makes this Earth a miserable place; Heaven is an Holy and so an Happy place." No further specificity is necessary. To these New Englanders, a heaven peopled with good friends and blazing hearths was more remote than a vacation to Fiji.

The First Great Awakening, in the middle of the eighteenth century, was a wave of heightened religious fervor that moved across the Atlantic from Britain to New England. It was triggered by such itinerant preachers as the Englishman George Whitefield, who taught that *anyone* could have a personal relationship with Jesus and that scriptural truth was accessible to anyone who could read or listen to the word; at revivals, preachers stressed conversion—their prayers for salvation were greeted with an emotional, enthusiastic response. In keeping with this populist message, heaven grew nearer. Preachers began to speak of an ever-after of sweet reunions, where, in the words of the New England Puritan divine Jonathan Edwards—better known for such harsh sermons as "Sinners in the Hands of an Angry God"—"there is the enjoyment of persons loved, without ever parting. Where those persons, who appear so lovely in this world, will really be inexpressibly more lovely and full of love to us."

THE SWEDISH VISIONARY

In *Heaven: A History*, Colleen McDannell and Bernhard Lang argue that some of the roots of the modern-day heaven, in which people imagine themselves living in domestic bliss with deceased relatives and friends, can be found in the writings of the eighteenth-century Swedish mystic Emanuel Swedenborg. Their position is controversial in the academy. Harvard's David Hall, for one, believes that modern-day ideas about heaven mostly have other roots, especially in an urge among Americans as the country matured to cast off the rigid austerity of Puritan values. It is fair enough to say, though, that Swedenborg was an important influence on some important Americans—and that his vividly embroidered heaven marks an early articulation of a view you hear a lot today. Stephen Prothero, a religion professor at Boston University, told me that Swedenborg was the eighteenth century's Oprah. He opened the door to what we would call New Age thinking and to the belief that heaven is just like this world, only better.

Jonathan Rose is a seventh-generation Swedenborgian, which means his ancestors were among America's first converts to the religion in the early part of the nineteenth century. He grew up in small, insular enclaves of Swedenborgians in the United Kingdom, Canada, and, finally, the woods and dales of Bryn Athyn, Pennsylvania, at a time when not taking another Swedenborgian to the school dance meant getting detention. He still lives in Bryn Athyn. Tall and rail thin, with a face you might call "sensitive," Rose is one of perhaps twenty thousand Swedenborgians in the world, five thousand of them in the United States. He is a minister and a scholar, who sometimes preaches in Bryn Athyn's Swedenborgian cathedral on Sundays and has devoted his life to translating Swedenborg's voluminous works from Latin into readable English. Rose can see how odd it is to be committed to promoting the religious vision of a person few have ever heard of—he understands that he lives in a bubble. Rose also knows that he is a lucky man. The Scottish-born American industrialist John Pitcairn (1841–1916), a devoted Swedenborgian, left a fortune to his church. Until 2001, Rose was a professor of religion at Bryn Athyn College, a tiny Swedenborgian school with 150 students and an endowment of $200 million. Today, as the editor of a new translation of the complete works of Swedenborg, he works out of a grand, Elizabethan-style stone-and-wood-beam office in a mansion that one of the Pitcairn sons formerly called home.

Emanuel Swedenborg was born in 1688, in Stockholm, into a wealthy family; his father was a Lutheran bishop. Swedenborg himself became a scientist and an engineer, and had little interest in religion until he was nearly sixty, when he began to have visions and hear the voices of angels—voices that didn't stop until he died, twenty-seven years later. His first waking visionary experience occurred in 1745. He was dining alone in London, he said, when he saw in a corner a shadowy figure who he soon realized was Jesus. "Eat not so much," the Savior said. After that initial vision, Swedenborg heard the angels all the time: while sleeping, while awake, while attending dinner parties at people's houses. And the angels took him to heaven and showed him around. Swedenborg believed that in heaven, people became angels—not physical in themselves but capable of enjoying physical pleasure, including food and sex. "Heaven is indistinguishable from this world to the degree that people don't realize they've died," says Rose. In heaven, Swedenborg informs us, angels had homes, "like the dwellings on earth which we call homes, except that they are more beautiful. They have rooms, suites, and bedrooms, all in abundance. They have courtyards, and are surrounded by gardens, flower beds, and lawns."

Swedenborg's heaven is a place of diverse characters, architecture, sights, and sounds. "It sounds like Manhattan," says Rose. It's "a work- and use-oriented culture." Angels are "all everywhere, everyone's mixing together."

Before arriving in heaven, people are sorted according to their likes and dislikes in what Rose calls "a melting-pot area." Swedenborg re- counts meeting Martin Luther there, where they argue over Luther's ideas—and Luther revises and repudiates his Reformation theology. In heaven, Swedenborg meets Cicero, who wrote that good Roman citizens would, after death, ascend to heaven where they'd meet other citizens, as well as loved ones. The whole time he and the great man are talking, Rose says, laughing, "they're being hassled by these Chris- tian spirits." Swedenborg was frustrated and embarrassed by the Chris- tians, "the way I would feel," explains Rose, "being in France and being embarrassed by other American tourists."

Swedenborg believed that marriages continued in heaven, that the spirits of the dead existed near the living in a kind of parallel uni- verse. Rose lost his first wife to cancer when he was thirty-eight years old, and was gravely disappointed when he did not feel her presence as often or as strongly as his religion led him to expect that he would. "Swedenborg says where there's true love, the sense of communion is not broken by the death of one," says Rose. And yet when Anne died, "she kind of disappeared. That was a shock to me." Her vanishing from his life didn't lead Rose to question his faith or his love for her, but he did occasionally worry that "we must not have been that close or I would have had that experience." Finally, years after her death, he had a dream of playing music with Anne—something they did together in life—that gave him great comfort and relief. And as he continued to study Swedenborg's works, he realized that he had made a mistake: Swedenborg, he now believes, never says or implies that a lack of contact between the living and the dead bears any relation to the quality of a marriage. Rose has since happily remarried and is, for the time being, holding questions of true compatibility in abeyance. "I've left who ends up with whom in the Lord's hands."

Swedenborg believed, as the Mormons do, that people in heaven live together in large communities in a place that resembles a better earth. But unlike the Latter-day Saints, the Swedish visionary believed that the ties that bind are not those of blood but of like-mindedness. (Rose regards the Internet, which organizes the world according to intersecting interests, as something like a Swedenborgian heaven-on-earth.) Discussing Swedenborg's celestial affinity groups in his office, Rose's voice becomes thick with emotion. "He talks about people being able to recognize each other, like a spiritual family resemblance," Rose says. "The people you see there, it's beautiful—it's like people you've known all your life, best friends forever." In our earthly lives, he adds, we spend so much time with people who are unlike us that we continually have to be on our guard; just imagine a life of perfect comfort and safety, in which we completely love and trust all those around us.

The Swedenborgians claim that their founder's influence is far-reaching. The theologian Henry James Sr., the father of William and Henry James, they note, was an avid Swedenborgian. Helen Keller, who wrote glowingly of Swedenborgism in her 1927 book *My Religion*, once reportedly declared: "Swedenborg does such good to me that I long to scatter his teachings to men and women wherever I go." In an essay published in 1850, Ralph Waldo Emerson called the Swedish mystic a "sublime genius." Religious crazes come and go in America. I decline to decide which one is the craziest; longevity and numbers of adherents do not seem to me adequate measures of the truth of a religious movement. In that light, the way in which Rose clings to his tradition—as if he were the last Gullah speaker in Georgia—seems to me both admirable and bittersweet.

By the early 1800s, then, Americans were ready for a heaven that was "closer" to the earth, more intimate and more consoling. This "humanizing" of heaven, says Harvard's Hall, went hand in hand with the Second Great Awakening—an evangelistic fervor that gripped the nation—and other civil rights ambitions of the period: a loathing of public floggings and executions, of corporal punishment in schools, and, of course, of slavery. Reformers such as Lyman Beecher and Charles Grandison Finney preached particularly against Calvin's doctrine of predestination. They assured Christians that their efforts on behalf of their fellow creatures mattered to God in heaven. Women, to whom the Awakening provided new opportunities for missionary work and evangelization, railed in particular that a grieving mother might have to consign a beloved and innocent child to a cold, impersonal, and terrifying heaven.

In 1832, Henry Wadsworth Longfellow wrote "The Reaper and the Flowers," a poem meant to relieve a mother's grief. In it, a reaper comes to earth and with tears in his eyes picks the most beautiful flowers and carries them away. The reaper is, in fact, an angel; the flowers are the souls of dead children; the bouquet is, of course, for Jesus. "And the mother gave, in tears and pain, / The flowers she most did love; / She knew she should find them all again / In the fields of light above."

Over the next few decades, the idea of heaven as a place for family reunions blossomed. Joseph Smith, the founder of the LDS church, believed that at death people were ushered into heaven not by an angel but by their family who had arrived before them. "I actually saw men," said Smith in 1842 in a funeral sermon, "I actually saw men, before they had ascended from the tomb, as though they were getting up slowly. They took each other by the hand and said to each other, 'My father, my son, my mother, my daughter, my brother, my sister.' And when the voice calls for the dead to arise, suppose I am laid by the side of my father, what would be the first joy of my heart? To meet my father, my mother, my brother, my sister; and when they are by my side, I embrace them and them me." When Jesus comes again, Smith preached in a sermon the following year, people will rise with their physical bodies—and with their social relationships. "The same sociality which exists among us here will exist among us there, only it will be coupled with eternal glory."

HEAVEN DRAWS NEAR

Six hundred and twenty thousand young men died in the Civil War— 2 percent of the American population at the time. These statistics reflect many multiples of grief. Wives, mothers, daughters, sisters, and sweethearts mourned over the devastation of their futures, and the heaven being offered from most pulpits—austere, brilliantly lit, full of God but devoid of humanity—offered no comfort. It was then, Colleen McDannell and Bernhard Lang argue in *Heaven: A History*, that the idea of a domesticated heaven really took hold in the popular imagination. In 1868, the twenty-four-year-old Elizabeth Stuart Phelps published *The Gates Ajar*, a novel that became the *Da Vinci Code* of its time, denounced as heresy by some and heralded as revelation in others. It tells of a young woman who is left alone in the world by the death of her brother Roy in the war. "Shot dead," the telegram read. The heroine, Mary, descends into a hell of grief. She won't see people, she feels shut off from God. The dour old village pastor chides her but offers no consolation.

And then she receives a visit from an aunt, whom she had not seen since she was a girl. Aunt Winifred turns out to be like a fairy god-mother or, more exactly, an angel. Winifred is learned in the Bible and devotional literature but she is also an expert in grief, having lost her husband prematurely to illness. Together with her young daughter, Faith, she brings Mary a Christian vision of heaven full of earthly de-lights and human company. Heaven, she teaches, is full of mountains "as we see them at sunset," trees "as they look when the wind coos through them on a June afternoon," pretty homes full of flowers and friends. Quoting the British romantic poet Charles Lamb, Winifred tells Mary that in heaven shall be "summer holidays, and the green-ness of fields and the delicious juices of meats and fish, and society . . . and candle-light and fireside conversations, and innocent vanities."

Roy, Winifred tells her niece, is nearby all the time, "trying to speak to you through the blessed sunshine and the flowers, trying to help you and sure to love you." When the time comes for Mary finally to join Roy in heaven, she will listen to his jokes, "see the sparkle in his eye, and listen to his laughing voice." She will be able to touch him and kiss him. When Mary hesitates, as she does from time to time, to ask if Winifred's teachings are heresy, Winifred simply laughs. *The Gates Ajar* sold eighty thousand copies in the United States by the end of the century, surpassed in sales only by the Bible and Harriet Beecher Stowe's *Uncle Tom's Cabin.*

As heaven grew "near," Americans began to look beyond the bound-aries of conventional Christianity for assurance that they would see their loved ones again. Some became Universalists—so called because they believe that everyone attains heaven in the end. Universalism started in Europe also in opposition to Calvin's doctrine of predestination and flourished in the mid-1800s. While its adherents were aligned with the reforming social goals of Finney, Beecher, and other leaders of the Second Great Awakening, they rejected the notion preached at so many revivals that salvation was for Christians only. By their own definition, Universalists are Christians who "don't believe a loving God could condemn anyone to hell for eternity." In 1961, the Universalists formally joined forces with Unitarians (who don't believe in a Trinitarian God), creating the Unitarian-Universalist denomination.

Others—many of whom were also Universalists—began to talk to the dead. Spiritualism started as a vogue around 1850 and peaked during the Civil War, despite continued disapproval from religious authorities. Among the earliest and most famous Spiritualists were Margaret and Kate Fox, two young sisters from Rochester, New York, who claimed the dead spoke to them through a series of knocks, raps, and taps. Their gift was discovered one evening when their mother heard a strange tapping on the walls of their small cottage and wondered aloud if the house was haunted. When the tapping continued, the girls began to ask questions of the spirit and soon it told them—through more taps—that it was the ghost of a Jewish peddler who had been murdered, beheaded, and buried under the house. The Foxes told their neighbors, who arrived in throngs, and the sisters became instantly famous. One night, according to contemporary reports, hundreds of people crowded into their cottage to talk to the dead.

Soon the girls took their show on the road, contacting spirits in theaters and holding séances in private homes. In New York City, Representative Horace Greeley expressed his conviction that the Fox sisters' gifts were genuine. "It would be the basest cowardice not to say that we are convinced beyond a doubt of their perfect integrity and good faith," he said. On his deathbed, the writer James Fenimore Cooper reportedly said, "Tell the Fox girls they have prepared me for this very hour." For thirty years, the Fox sisters traveled America and Europe, charging admission to people interested in talking to spirits and inspiring generations of imitators. Spiritualists, whom today we would call "channelers," rejected conventional ideas of heaven and hell, and believed that everyone in the end would be saved.

One astonishing night in 1888, however, Margaret Fox repudiated Spiritualism altogether. Before a crowded hall at the Academy of Music in New York City, she vowed to tell "the truth, the whole truth, and nothing but the truth." Then, according to a *New York Tribune* account of the event, she put her stocking feet on a sounding board—a wooden plank that amplifies sounds—and she popped the knuckles in her toes. "The audience heard a series of raps, 'rat-tat-tat-ta-tat-,' increasing in sound from faint to loud, and apparently traveling up the wall and along the roof of the Academy." Avowed Spiritualists, undeterred, said the Academy of Music performance should be discredited, but ultimately the sisters became a punch line. According to her 1893 *New York Times* obituary, Margaret died "in poverty and obscurity."

Especially during the Civil War period, interest in Spiritualism was intense and widespread. Mary Todd Lincoln became obsessed with Spiritualism, and held séances at the White House at least eight times, eager to see her sons Eddie and Willie, who had died from childhood illnesses. (It is uncertain how often President Lincoln attended or how seriously he took them.) Mrs. Lincoln, who was able to get into trances herself, once declared to her half sister: "Willie lives. He comes to me every night and stands at the foot of the bed with the same adorable smile he always has had. Little Eddie is sometimes with him." Planchettes, the triangular pieces that move across Ouija boards, went into mass production during the war and the first national Spiritualist convention was held in 1864.

But by the beginning of the twentieth century, the pendulum swung back again—and a new religious elite, influenced by Darwin, Marx, and Freud, impatient with a religiosity they saw as too naive and unsophisticated, once again emptied heaven of its people, its villages, its rooms and gardens. In exchange they offered an idea of the afterlife that was intellectually demanding and emotionally void. The twentieth-century Protestant theologian Paul Tillich (1886–1965) called the reliance on "real" images of heaven a "literalistic distortion" with "neurotic consequences." And so across the country, mainline pastors talked about heaven only when they had to—and without making it seem *too* real. "I'm not interested in speculating on the architecture or the geography" of heaven, the Methodist minister J. Philip Wogman told *Time* magazine in 1997. "We dig a lot deeper. I preach on trust in God."

But the fundamental human longing to be with the ones you love, especially family members, defies Tillich's derision and strict rationality, and online the idea that people will see lost loved ones has launched a cottage industry. Dozens of small companies help the bereaved to set up memorial sites, like Facebook pages, where friends and relatives can post their memories and condolences. Sammy Hicks, a retired electrician from Disputana, Virginia, died in February 2009 and his children put up a page on ChristianMemorials.com. "You were a great man daddy, and you will be missed so much. We will keep your memory alive through Joshua, he loved his Papaw so much. I'll see you in heaven." Johnnie Lane Parker Sr. died of congestive heart failure that same month in Hughesville, Maryland. On the site ValleyOfLife.com, his friend Tony Jr. wrote, "JP will be greatly missed, but his spirit will live on. We will see him again. Believe it." A smiley face punctuated that message.

Anne Foerst, a German-born believing Lutheran, is also a professor of computer science and theology at St. Bonaventure University in western New York. For years she ran the "God and Computers" project at MIT. She is interested in the way computers reflect and change our ideas about religious faith. Foerst believes that the ubiquity of online memorials reflects a failure on the part of mainline Protestant churches to give people a meaningful heaven. "Here come the intellectual Protestants and try to make theology something that is rational," she said to me one morning in one of our regular phone conversations. From the time we are about three years old, she explains, we have an understanding of being unlike other people, and we live our lives with an overwhelming yearning to overcome that primal estrangement. We strive constantly—in marriage, in work, in friendship—to close the gaps between ourselves and other people. Heaven as a place of happy reunions, especially with family members, is an imaginative way of resolving loneliness and the inevitable schisms between ourselves and the people we love. "We are close to our family, we are alike, but at the same time there's so much passion—even negative passion. Even if you fight with your siblings, a family member is someone you cannot lose." The Internet, says Foerst, finally creates for people "a space of unestranged togetherness." The memorials themselves may be kitschy or saccharine, but the longing that produces them is not.

Online, the question of whether people will enjoy the company of pets in heaven triggers explosions of sentiment. For a 2001 poll conducted by *ABC News* and BeliefNet.com, a website for adherents of all religious faiths and spiritual inclinations, 43 percent of people answered the question "Will we see our pets in heaven?" in the affirmative. Among pet owners, that number was 47 percent. And at Critters.com, pet owners who memorialize their dogs or cats or parrots often express the certainty that they'll meet again. To comfort bereaved pet owners, the site's Webmasters have posted a little prose-poem called "The Rainbow Bridge." This bridge is where deceased pets wait, restored to their former vigor and health, for their owners to come get them. They play happily together, in warmth and comfort until "the day comes when one suddenly stops and looks into the distance. . . . His eager body quivers. Suddenly he begins to run from the group, flying over the green grass, his legs carrying him faster and faster. You have been spotted. . . . You cling together in joyous reunion." The poem seems to have struck a chord: the site currently indexes hundreds of customized pages mourning lost dogs and cats, or paying poignant tribute to departed ponies, iguanas, and parrots.

Americans continue to believe that their dead wait for them nearby—on a rainbow bridge, if you will. Twenty-one percent of people believe you can talk to the dead, according to a 2005 Gallup poll. On a sleeting, gray evening, I called on Glenn Klausner. He is a medium with a good reputation, who came recommended by a friend. Klausner is in his mid-thirties and grew up in Brooklyn. His apartment is on an upscale block of town houses and apartment buildings on Manhattan's Upper West Side, but his own building is modest and a little in need of repair.

He works out of his small apartment, talking to the dead in his living room, notable for its lack of any personal detail. It has a leather couch, a couple of houseplants, and a sleek black table. Klausner's physical appearance is similarly neutral. Aside from a bushy head of glossy dark hair, cut to emulate a young Paul McCartney, he transmits no obvious sexuality, no idiosyncrasies. He is thin and pale, dressed in jeans and a sweatshirt. He has a tiny silver earring in his left ear.

Klausner has been talking to the spirits of the dead since he was a child—and he adds that he always knew who was on the phone before his mother answered it. He has been a professional medium since 1995. He works four days a week, and sees about forty clients a month, by phone and in person. Most of those who contact him have recently lost a loved one. He charges $350 an hour.

"Heaven," he says after we sit down at the black table and he hands me a glass of water, "actually is not above us or below us, but all around us. When I've asked spirits what it's like to transition out of the body, they say it's nothing—like walking from the kitchen to the bathroom." Spirits design their own personal heavens, he says, according to their own desires. Some live in mansions, some in high-rise apartments, some in country houses. They travel easily between one another's dwellings because they can circumnavigate the world in an instant: one spirit recently told Klausner that he had been to the Galápagos Islands; he was helping his granddaughter on earth plan a vacation.

I ask Klausner whether it's true that in heaven there is no sickness, anger, or jealousy, and he says yes. "There are no health care issues, no monetary issues—if someone had stage four cancer they don't take that with them." I ask what kind of bodies the spirits have, and Klausner answers this way: "They have what's called an astral body. Let's say you had an uncle who looked like Dick Van Dyke—I can see that. [I don't.] The spirits that I deal with, it's really their personality that comes through, though a lot of spirits go back to looking like their best. They'll come back looking like they did when they were thirty."

I can't help myself. I ask him if he can channel someone for me. But Klausner refuses to channel spirits by name. He simply contacts the spirit world and talks to whoever comes. I am hoping that he'll find my grandmother, Rita, with whom I found so much comfort as a child— and well into my thirties. She and my grandfather put me up in their extra bedroom when I first moved to New York, broke and ambitious, and over the years she took care of me in small, important ways. She took me with her on the express bus to Loehmann's discount clothing store in the Bronx. She cooked for me, she stroked my hair as I rested on her couch. We drank scotch together, and watched tennis on television. I miss her all the time, but especially around the High Holy Days—not because she was religious; we never went to temple together—but because she was such a skilled cook. For more than a decade, I spent Rosh Hashanah sitting at her little round dining table, just the three of us: my grandmother on my left, my grandfather across from me, and me, the oldest grandchild, holding forth or sullen, but always knowing how cherished I was. Flowered napkins were rolled into napkin rings and a silver tray always held matzo and crackers. My grandmother served gefilte fish and soup, brisket and couscous, salad with inimitable dressing—creamy, with shallots and fresh-cut dill— and poached plums for dessert. I have been telling my daughter how beautiful Rita was in the photographs of her as a girl, with long auburn braids down to her bottom and tied with ribbon. I would really love to see her.

The person Klausner finds on the other side is, unfortunately, unrecognizable. He's bald, Klausner reports, with fringes of hair on the sides of his head. He's a father figure, and he looks like Ed Asner. I don't know who that is, I say. My own father is, thankfully, alive— and he doesn't look like Ed Asner. My father-in-law? Klausner asks. Perhaps, I say. I never met my father-in-law, who died the summer before I met my husband, but based on the pictures I've seen, this doesn't sound like him. Klausner reassures me; sometimes, he says, spirits appear in forms different from the ones they inhabited in life. He listens to the spirit. My father-in-law is a brusque person, Klausner says; he has a temper. Klausner reports that the Ed Asner figure wants to apologize to my husband, Charlie, for not being a better father, for not listening to him. And though I'm fully aware that such an apology is emotional boilerplate—it could apply to any parent-child relationship—tears spring to my eyes anyway.

And then the trail goes cold. Klausner reports that my father-in-law was involved in some illegal activities: he hid money in his house. (My father-in-law was no criminal.) He reports that my father-in-law was a drinker of clear liquor. Wrong again: Laverne loved bourbon. He sends greetings from the afterworld from Alice, Roslyn, Dorothy, and Ruth. I don't recognize the names, and when I ask Charlie later, he confirms that they're no one he knows. At this point, I begin earnestly to try to bring Klausner back to his little apartment, but he's too far gone. The Ed Asner character wants to tell his wife that she's taking too much Valium. That's when the rest of my credulity vanishes. My mother-in-law is eighty-six years old. She's a little bit forgetful and not quite as quick on her feet as she used to be, but she doesn't take Valium and is as hale as a horse.

I will not rule out the possibility of an afterlife, but I do rule out the idea that my father-in-law, a man I never knew in life, is watching over me, disguised as Ed Asner, on some astral plane—and that Glenn Klausner has access to him. Klausner seems like a perfectly nice person and he probably believes his own stories. But I remain doubtful. In a phone call later, Klausner rejected any question about the authenticity of his gifts or the solace he has given to so many. "I have years of track record," he says. "I've done a lot of tremendous work. Not every reading is dead on. I'll be honest, I work with a person who's paying me, and if the reading's a bomb, I don't take a dime. . . . That shows integrity of character in me." For my part, I believe Klausner traffics in grief. If he really has this gift, he should always give it for free.

NINE

IS HEAVEN BORING?

The sound track to my college years was the band Talking Heads, and one of my favorite tunes at the time was their song "Heaven." David Byrne, their lead singer and songwriter, was an art-school refugee, a hollow-cheeked, dark-eyed artiste who crooned the song with a spaced-out sweetness. The refrain goes like this: "Heaven, heaven is a place, a place where nothing, nothing ever happens." Most of my friends from that era asked me to include that song in this book. Usually, they'd sing it when they asked me.

Like *Defending Your Life* and *The Lovely Bones*, the song "Heaven" is a cultural touchstone. I thought its popularity and resonance must mean something about what we believe about heaven now. Before meeting Byrne, I took a look again at those lyrics. The first verse describes heaven as a bar where the band is always playing your favorite tune. The second verse describes heaven as a party with no social anxiety, no chance of ever being ditched by your date in favor of someone more interesting or willing: "Everyone will leave at exactly the same time." But the final verse is my favorite. It describes a kiss, presumably a wonderful kiss, happening over and over in an endless loop.

> When this kiss is over it will start again,
> It will not be any different, it will be exactly the same.

It's hard to imagine that nothing at all
Could be so exciting, could be this much fun.

Byrne works out of a loft in New York City's SoHo district. He's now a visual artist as well as a performer, and his walls are covered with his photographs, but relics of his rock-star past are evident everywhere. The original painting for the album *Naked*—a smiling chimp against a red background in a gilt frame—hangs near the doorway to his office. A sketch for the album cover of *Speaking in Tongues* (gold, with Aztec-looking designs in the corners and designed by Byrne himself) lies on a desk in a plastic sleeve, waiting to be archived. Byrne has a deep, intellectual interest in religion, and as a purveyor of irony and disconnections, he continues to make art that turns Christian and other religious conventions on their heads. In 2001, he created a small book called *The New Sins*, which looks and feels exactly like one of Gideon's bibles—and which Byrne distributed anonymously in hotels and motels across the land, just as the Gideons did. *The New Sins* is an art project. Produced by Byrne, the avant-garde literary entrepreneur Dave Eggers, and their friends—all of whom collaborated under the name The Better-Emancipated Strivers for Heaven—it purports to disseminate wisdom of the scriptural kind. "Heaven and Hell are both metaphors," it says under the heading "What Happens When We Die?" "Good ones, too. So good, in fact, that they function as if they were real for many people, even unbelievers." A small foldout chart of hell relegates journalists to the eighth circle, wedged between market researchers and demographic testers.

I keep an image of Byrne in my mind, sweating and dancing like a madman in a humongous white suit in Jonathan Demme's Talking Heads concert film *Stop Making Sense*. As a result, everything about him in person is somewhat smaller than I expected. He's slender, his voice is quiet, and as he pauses to think about a question, he loses eye contact. He's wearing a knit cap, track pants, and sneakers, and though the famous cheekbones are still there, they're less jagged than they used to be.

We sit together in his small office and I ask him if he remembers writing "Heaven." "That was a rare one," he answers softly. "It was a very clear reasoning went into that song. The traditional Christian imagery we get handed is that no one is ever doing anything. People are lying around on clouds listening to a kind of boring harp music. Even in more Eastern concepts of Enlightenment or Nirvana, in those concepts it's also a thing where nothing really matters. You become disengaged and time stands still. The goal seems to be to attain a state where nothing happens. Nothing of any emotional import is ever going to happen and that's implied to be a good thing. When you put it literally like that, it doesn't sound as good.

"So I thought, 'Well, okay, I'll base this on my life at the time, going to a party, going to a bar or a club. It never ends, it goes over and over again. You kiss someone and it loops around and starts again. I'll write it as if I'm singing about something desirable and heavenly. I'll write about it in these terms that sound a little bit ridiculous. The voice singing it has an emotional longing. Everybody's like those bobblehead toys but totally blissed out about it.' "

I defend the song. When I was younger, I say, I would have agreed that it makes heaven ridiculous. Who would want to wake up every day and have everything be exactly the same? But now, encumbered and aging as I am, the endless repetition of pleasant activities sounds appealing. He looks back but doesn't say a word.

Do you believe in heaven? I ask him, finally. "Oh no," says Byrne, laughing. "I would like to believe in, like, karmic retribution or divine justice, some of which is implied by heaven. I would like to believe that the people who cut in line will get their just deserts. But I don't think they will."

With his song, though, Byrne has put his finger on the most nagging of all modern heaven questions: What's so great about heaven? In spite of war and a bad economy, twenty-first-century America *is* heaven compared to the world of the Bible and the Qur'an—and indeed compared to most places on earth where people live. Scholars suggest that our halfhearted credulity about the ancient promises of heaven is rooted in part in our contemporary material comfort. "No one demands that we have an insight or an inspiring word," writes Peter Hawkins in his 2009 book *Undiscovered Country: Imagining the World to Come.* "It is enough to be spiritual, not religious—to shrug our shoulders and say, good naturedly, 'Frankly, I have no idea.' " Pastors don't preach very often about heaven anymore, and even modern hymns—the locus of some of Christianity's most salient images— "tend to focus more on the person of Jesus, on salvation, and just on general praise to God rather than the afterlife or heaven," says David Music, professor of church music at Baylor University.

Real life has now delivered on so many of Scripture's promises of heaven. We have glittering cities, blossoming public parks with gushing fountains, libraries filled with books in every language, and a government predicated on justice and equality for all. We have gyms and Pilates classes that keep our bodies youthful, Botox to ensure that our faces remain unlined. At the peak of the housing bubble nearly 70 percent of Americans owned their own homes; even in the foreclosure crisis, the "many mansions" that Jesus promised are a reality here. Our supermarkets sell ripe fruits in and out of season—plus milk by the gallon and honey made by orange-blossom- or clover-loving bees. They even sell wine and beer that won't get you drunk, just as the Qur'an advertises. To put it very crassly, it's hard to convince people of a bountiful hereafter when in life they can buy almost anything they want.

The concern over celestial tedium is not entirely new: it can be traced back to the dawn of the industrial revolution, when Americans began to think of their land as a prosperous place where anyone could—and did—ascend to greatness. "Singing hymns and waving palm branches through all eternity is pretty when you hear about it in the pulpit," complained Mark Twain in 1909 in *Captain Stormfield's Visit to Heaven*, "but it's as poor a way to put in valuable time as a body could contrive." Modern Jews suffer especially acutely from the absence of convincing or exciting images of the afterlife. Rabbi David Fohrman, scholar in residence at the Hoffberger Foundation for Torah Studies, put it this way in a 2006 essay. "Say, for example, you really enjoy cruises. Someone comes along and offers you a free cruise to Alaska. . . . What if that cruise lasted for eternity? Year after year, it's the same icebergs and the same old Beluga whales."

The heaven-is-boring problem has theological as well as sociological roots. The heaven preached especially by twentieth-century mainline ministers was devoid of the cinematic particulars that made the visions of Dante and even Elizabeth Stuart Phelps so tantalizing. Too much imagining of heaven was seen in mainstream Protestant circles as the sign of an unsophisticated faith. Barack Obama's former preacher, the Reverend Jeremiah Wright, complained about this in a 1990 sermon at his Chicago church. His "educated friends," he said, wished he wouldn't talk so much about heaven "because that's so primitive, you see." These friends, Wright said, believed heaven-talk detracts from the real message of the Gospels, which is justice. But heaven, Wright insisted, is at the center of Christian belief. "If I drop heaven, I'm going to lose the first verse in my Bible," Wright preached. "If I drop heaven, I'm going to lose two of my Ten Commandments. . . .

If I drop heaven, I'm going to have to stop praying my favorite prayer, 'Our Father.' . . . If I drop heaven, I'm going to have to do away with the Second Coming; I'm going to have to get rid of Pentecost. I'm going to have to throw Revelation out of my Bible. . . . Don't make me drop heaven!"

Martin Marty, a longtime friend of Wright's, is the premier historian of American religion, having taught at the University of Chicago for thirty-five years and written numerous books, including the best-selling *Pilgrims in Their Own Land* (1984). Now eight-two, he lives in a modern high-rise apartment building on the Loop with his beloved wife, Harriet. Another, smaller apartment on a lower floor serves as his office. Growing up—he is a Lutheran from Nebraska—he rarely remembers hearing his minister talk about heaven. The one image that sticks in his mind is this: "He would say, 'There's a mountain somewhere a mile high, and a mile long, and a mile broad.'" (Marty adds that there weren't too many mountains in Nebraska, but he'd seen pictures.) The preacher would continue: "He said, 'Every hundred years, a bird pecks a mouthful from the mountain and flies away. As soon as that mountain has disappeared, one second of eternity has gone by.' I said, 'I don't want that—it's a pretty boring thing.'" The heaven offered in early- and mid-twentieth-century pulpits, says Marty, was dull, dull, dull.

This emptiness in the place where heaven used to be created an opening in our imagination for every kind of satirist. Boring-heaven jokes are now as much a part of our culture as heaven itself. Robert Mankoff, the *New Yorker*'s longtime cartoon editor, told me one afternoon in his office that heaven is second only to the desert island as a favorite setting for his cartoonists. "Basically, we laugh at incongruity," he explained. "Why are people in clouds? Maybe an angel is agoraphobic. Maybe they don't have harps—they have iPods or boom boxes." The Simpsons, the hilarious cartoon characters who have become even more than the Archie Bunkers enduring American archetypes, worry frequently over the question of what they'll do in heaven. In one episode, Bart and Homer threaten to convert to Roman Catholicism while Marge frets that her WASP heaven will be much more boring than theirs—as indeed it is. When she goes up to take a look around, Marge finds herself trapped with sour-faced J. Crew–wearing badminton players, while Bart and Homer rage like rowdy wedding guests at the Mexican fiesta and the Irish bar in theirs. The British comedy troupe Monty Python took on heaven in their 1983 movie *The Meaning of Life*. This version of paradise is a Vegas-style nightclub, starring an eternally bland and upbeat lounge singer. His signature number—"It's Christmas in Heaven"—includes such lyrics as "It's nice and warm and everyone / looks smart and wears a tie."

For the devout, this sort of humor is nothing less than a crisis. Those who take seriously the promise of heaven clashed recently with those who don't over, of all things, a paper cup. Starbucks, the Seattle-based coffee company, prints thought-provoking aphorisms on its cups, and in 2007 it released this one: "Heaven is totally overrated. It seems boring. Clouds, listening to people play the harp. It should be somewhere you can't wait to go, like a luxury hotel. Maybe blue skies and soft music were enough to keep people in line in the 17th century, but heaven has to step it up a bit. They basically are getting by because they only have to be better than hell." The author of this quotation was Joel Stein, the young columnist for the *Los Angeles Times*, erstwhile contributor to *Time* magazine and generally acknowledged enfant terrible. Conservative Christian groups reacted strongly, not so much against Stein, but against Starbucks—a few even called for a boycott. Five people sent Stein a copy of Randy Alcorn's book *Heaven*.

Alcorn is an evangelical Christian minister who lives near Portland, Oregon, and specializes in heaven. He was sitting with his mother and reading to her from the book of Revelation as she lay dying of cancer more than twenty years ago when it hit him: heaven is a real place. Alcorn had gone to a Christian college and to seminary, yet "the impression I had gotten along the way was that heaven is a disembodied spirit realm where you drift around." After his mother died, Alcorn went back to Scripture and mined it for its descriptions of the afterlife. This research resulted eventually in his book. "I didn't come up with this on my own," he says. "It is my understanding of what the Bible is saying." Heaven will be on a renewed earth. "We'll be doing things. There will be places to go, people to see. . . . Scripture talks about the New Jerusalem. Heaven will be the New Portland, the New Boston." In heaven there will be a river, Alcorn says, and glorious parks, museums, and sports arenas. The heavenly cities will have no litter, no crime. Alcorn published *Heaven* in 2004. He doesn't have a church or a congregation: through his nonprofit Eternal Life Ministries, Alcorn evangelizes about heaven.

Virtually unknown in secular circles, *Heaven* is a must-read for committed evangelical Christians. Rick Warren, the charismatic and powerful pastor of Saddleback Church in Orange County, California—a man with millions of people on his e-mail list—told me to read the book nearly five years ago, and when I got myself a copy, I found that Warren himself had written a blurb for the front cover: "This is the best book on Heaven I've ever read." Without concrete visions of heaven, Alcorn believes, Christianity has no hope. After receiving Alcorn's books in the mail, Stein called him up and they had a polite conversation. Stein wrote a column about Alcorn, and Alcorn blogged about Stein online. Each was civil in print. Stein describes in his column speaking to a woman named Shelly Migliaccio, one of those who sent him *Heaven*. In her vision of heaven "the colors are more brilliant, we all have jobs we love, we are free of the lies and horrible stuff she sees on the news. And for at least a little while . . . I believed in Migliaccio's heaven too."

THE RETURN OF HEAVEN

The fight to bring back vivid images of heaven that resonate in today's world is occurring on two fronts. On the one hand, you have the conservative faithful, whom I will discuss further on. On the other, you have the "seekers"—the approximately 30 percent of people who call themselves "spiritual but not religious." They cobble together a picture of heaven based not on creeds and dogma but on what they've seen in movies and on television and what they've read in books, along with residual messages from parents, grandparents, and Sunday school lessons—and, especially, on their own, individual experiences of transcendence. They think of heaven as a walk in the woods, a trip to the art museum, a night at the symphony, or an evening out with friends. These people tend to imagine heaven as the perfect culmination of their greatest desires—no restrictions.

For these folks, heaven is whatever you want it to be. "To me," wrote Ernest Hemingway in a 1925 letter to F. Scott Fitzgerald, "heaven would be a big bull ring with me holding two barrera seats and a trout stream outside that no one else was allowed to fish in and two lovely houses in the town; one where I would have my wife and children and be monogamous and love them truly and well and the other where I would have my nine beautiful mistresses on 9 different floors." In a song with the same title, the country singer Hank Williams Jr. proclaimed that "if heaven ain't a lot like Dixie," he doesn't want to go there. These images may be ironic, even humorous, but they make the point. For many Americans, heaven is the kingdom of ultimate personal fulfillment.

A striking example of this individualistic vision is *The Lovely Bones*, which spent seventy-eight weeks on the *New York Times* best-seller list. Five million copies are in print. The book's wild success was due in part to its lurid subject matter (the narrator, fourteen-year-old Susie Salmon, is monstrously murdered) and in part to its detailed, personal vision of heaven. In *The Lovely Bones*, everyone lives in a personalized heaven, and these heavens overlap with each other. Shot putters and javelin throwers play in the fields of Susie's heaven, forming a kind of sporty backdrop, but when the sun goes down, the athletes retire to their own heavens. Susie has a best friend—they chat as they eat peppermint ice cream—but the friend leaves Susie for hours at a time to go play her saxophone in a jazz band. Susie sits in a gazebo in her heaven; there—like the medieval visionaries—she can see the whole earth from a distance, as if in an airplane. At the same time, she can watch her friends and family members up close, as if she were in the room, as they fall apart and, finally, pull themselves back together.

Though the book of Revelation says there will be no tears in heaven, Susie has not left her emotions behind. She is not, as Byrne puts it, "blissed out." She feels vengeful rage at her murderer, as well as typical adolescent loneliness and isolation. She regrets her inexperience with sex and watches from heaven as the boy she loves gets over her. She lives in heaven, in other words, much as she would have lived on earth. In the end Susie is finally able to "move on" to another sphere where she will be able to let go of her earthly concerns. Christians have noticed, rightly, that God was nowhere in the picture, and suspected, again rightly, that *The Lovely Bones* implied an indictment of traditional Christian views of heaven. "When there is more talk of heaven in novels, television shows and pop songs than in sermons," wrote Mark Ralls, pastor of St. Timothy United Methodist Church in Brevard, North Carolina, in the *Christian Century*, "Christians must shoulder some of the blame for the fact that visions of life beyond death fail to include God."

I once interviewed Alice Sebold and she conceded the point. She grew up Episcopalian, and Susie's heaven is a rejection of what Sebold herself learned in church. "I've always felt that there were so many rules and exclusions out there in conceptions of redemption and the afterlife—it didn't include me and a lot of my friends. The guiding principle in [my conception of heaven] is that it's inclusive: it allows you to have what you want and what you desire." Sebold doesn't want heaven to be a perfect place: "There needs to be a little angst in heaven to make me happy."

In a universe where all people received exactly what they want in heaven, how do they manage to coexist? One person loves dogs but another hates them. How does this work out in heaven? The undergraduates in Stephen Prothero's Introduction to Religion class at Boston University confronted this problem squarely in 2007 when Prothero asked them to divide into small groups, design a religion, and then vote on the best one.

Prothero is the kind of teacher who, I wrote in a *Newsweek* profile, "makes you want to go back to college." Tall, blondish, and rangy with an occasional scrubby beard, Prothero looks like the young Billy Graham if Graham were a world religions professor and not an international evangelist. Prothero's lectures make connections designed to delight the postadolescent mind. On the day I observed one of his classes, he veered with agility between Hinduism, Docetism, Socrates, and Disney World. His main complaint about his students is their fervent political correctness. They refuse to articulate any belief that might offend. In an environment as diverse as BU, this political correctness works to suppress what Prothero believes may be the most interesting, if provocative, ideas. Talking about religious truth in general—and heaven in particular—can be a minefield.

The winning religion was designed to please. It was called ♩-ism. (Prothero and his students pronounced it "huh-ism.") Its prophet is Tupac Shakur, the rap star who was shot and killed in a gang fight in 1996. According to ♩-ism, Tupac became enlightened in the world beyond; he evolved into a magnificent musical savant, the celestial embodiment of all music, all rhythm, and all dance. Human life is called "The Party": people dance and feel the rhythm throughout their lives, and they attend weekly services in "Music Domes," where there are no ministers but deejays. After death, adherents ascend to "The After Party" (unless they've been bad, in which case they go to a place of eternal silence, called "The Hangover"). At "The After Party," individuals live out all their earthly dreams. They can skydive or dine at five-star restaurants. As in Sebold's heaven, no one's dreams impinge on anyone else's.

In the heaven of ♩-ism, people can have bodies or not, as they wish, and there is no God. "It's not a traditional afterlife," concedes Adam Greenfield, who was one of the religion's founders. "It's more like a modernized afterlife."

Among young people, "the afterlife is an afterthought," says Prothero, ruefully. Prothero himself was raised Episcopalian and had a brief flirtation with evangelical Christianity. When he was a young man, heaven, he says, was not just something he believed in; it was someplace he really wanted to go. Like so many people I've met— and like me—he mourns the disappearance of traditional heaven. His 2007 book, *Religious Literacy*, is an explication of that disappointment. Too many people, he writes, don't know the Scripture of their own tra-

dition—let alone that of their neighbors. They don't know what they
believe about God and heaven and are afraid to hash their ideas out
in public for fear of causing offense.

To put things right, the orthodox—Jews, Christians, and Muslims—
are trying to revive belief in a traditional heaven by embracing its
mystery. We don't know exactly what it will be like, they say, but it
will be wonderful beyond our imaginings—not boring, certainly not
ridiculous. In two books, *Paradise Mislaid* and *A History of Heaven:
The Singing Silence*, the theologian Jeffrey Burton Russell attempts to
re-create a picture of heaven and heavenly activity that both adheres
to a traditional Christian view and appeals to modern readers. "My so-
lution," Russell told me, is to "get a definition that isn't hard and fast;
it's fuzzy bordered, but it reflects the tradition over the past twenty
centuries." His heaven is radically—and you could even say euphor-
ically—paradoxical. The saved are there *and* everyone is there. It is
both in time *and* out of time: "It is motion *and* stillness; it is silence
and song," as he writes in *A History of Heaven*. It is, above all, God's
love. "So those whom we have never known, or with whom we have
quarreled, even those whom we have hated or who have hated us, all
these we love, for whatever is evil in them has been washed away, and
all that remains is the pure goodness that God has made, which is perfect love loving." The sillinesses of contemporary visions—ice cream
and puppies—have no place here.

What is so wrong, modern theologians ask, with believing with all
one's heart in a heaven that's both known and unknown? In his recent
book *Surprised by Hope*—a play on C. S. Lewis's *Surprised by Joy*—the
Anglican theologian N. T. Wright exhorts Christian readers (and, as
he says, "anybody else who's listening!") to forget about the fuzzy cultural concept of "heaven" and reclaim something more like the early
Christian view of the afterlife. It's a two-stage reality, he says. God's

final new creation will come at the end of everything. It is "like this present world only much more so, its physicality more real, its beauty more vivid, its pulsating life more intense and at the same time its peace more deep and rich." In the meantime, Christians should live with the knowledge that since the resurrection of Jesus, "God's sovereign reality has come to new birth within our present world." What this means, on a day-to-day level, is that believers should keep the heavenly ideals of beauty and justice continually in mind. Heaven is hope—but *real* hope—"for change, rescue, transformation, new possibilities." The Orthodox Jewish scholar Jon Levenson is, as we've seen in chapter 5, committed to bringing a full view of resurrection and the afterlife back to contemporary Jews. Such a view, he believes, can only be attained through regular practice and prayer. Heaven, he told me in an interview, "should be thought of as the perfection of everything in this world and then some." The UCLA jurist Khaled Abou El Fadl offers an interpretation of the Qur'anic heaven that's rational and mystical at the same time. Heaven is light, he tells me, but it's also enlightenment. "Our true tranquility is to know God, which is to know yourself. First God gets our attention with all this material stuff, then after that it's explorations of the intellect and the spirit."

WE WILL SEE GOD

Byrne, Sebold, Mitch Albom, and the students at Boston University are all missing the main point, the orthodox say: in heaven, *we will see God*. Saint Paul promises that while on earth we see through a glass darkly, in heaven we will see God face-to-face. The first Epistle of John promises that, in heaven, "we will see him as he is."

But what is it like to see God? The Church Fathers, the writers and thinkers who in the first few centuries after Christ wrote the theology that would form the foundations of the Roman Catholic Church, tried to work this out. Would they see God with their dead, human eyes? With their resurrected eyes? Or would they see God, somehow, with their spirit—a kind of transcendent physical vision? "Perhaps," Augustine wrote, "God will be known to us and visible to us in the sense that he will be spiritually perceived by each of us in each one of us, perceived in one another, perceived by each in himself." Perhaps, in other words, in heaven God is everywhere and visible in and to everyone.

Dante, as we've seen, imagined seeing God as gazing at a very bright, all-encompassing light, a light that embodies and bestows love and perfect understanding. "For my sight, becoming pure / rose higher and higher through the ray / of the exalted light that in itself is true." With this imagery, Dante was giving poetic form to the theology of Thomas Aquinas, the medieval monk who insisted that knowing and seeing God was the most any human could hope for. Our ultimate destiny, Aquinas held, was "perfect union of the soul with God, insofar as it enjoys him perfectly, seeing him and loving him in perfection." As if anticipating David Byrne's complaint, Aquinas adds that "nothing that is contemplated with wonder can be tiresome."

But for other Christians, then as now, such a vision was too abstract. In the medieval period, while Dante was framing a heaven of pure light and perfect understanding, religious mystics took the idea of perfect union with God in another direction altogether. Being with God is sublime, they said, casting about for metaphors. It's like sex without intercourse, inebriation without alcohol, warmth without fire. Bernard of Clairvaux, the founder of the influential Cistercian abbey at Clairvaux, imagined union with God to be something like the heady feeling of having had too much wine. "In the life of immortality," he wrote in the twelfth century, "we will be inebriated and overflow in wonderful abundance. And the bridegroom intoxicates his friends. Hence results inebriate sobriety, which is inebriated by truth, not by wine; which is soaked not in wine, but burns with God in eternity." The thirteenth-century German nun Gertrude of Helfta imagined her union with Christ as sexual—but also somehow completely chaste. In one of her visions, which she wrote down in the third person, "the Lord took her into his arms, holding her fast in his embraces and caressing her tenderly. . . . He covered her eyes, ears, mouth, heart, hands, and feet with kisses." She hoped that when she died she would achieve a perfect union with God in heaven. "In conjugal love and nuptial embraces show me your greatness . . . in a kiss of your honeyed mouth take me as your possession into the bridal chamber of your beautiful love." Her communion with God was individual—and yet not. Every virgin committed to Christ could experience heaven in this way.

In Judaism, God has no body, but Jewish mystics have long imagined life after death as a union with God; they have sometimes described that union in physical terms. Merkabah mystics, in the Second Temple period, reached ecstatic states by meditating on the vision of the chariot ascending in Ezekiel 1: "Like the bow in a cloud on a rainy day, such was the appearance of the splendor all around. This was the appearance of the likeness of the glory of the LORD." The goal, according to Merkabah literature, was to "gaze upon the King in His Beauty." The following account, attributed to the first-century rabbi Ishmael and edited into its present form by the eighth century CE, demonstrates God's—and heaven's—unimaginable hugeness. "The soles of his feet fill the whole world. The height of each sole is three ten thousands of parsangs [roughly ninety thousand miles]." The Zohar, a Jewish text written in Spain in the late medieval period, promises that individuals will upon their deathbeds be granted a vision of the Shekhina, a formless radiant presence that comes from God. In fact, the soul does not leave the body, the Zohar says, until it sees the Shekhina. Righteous souls will then cleave to it "in joy and love." The unrighteous will be left in the cold, "like a cat which is driven away from the fire."

Sufi mystics talk about seeing God, but obliquely, in the ritual they call Sema—known to us as the whirling of the legendary dervishes. In the Sema, holy men wearing white robes and tall hats gather in a circle. As the circle turns, the monks also whirl individually, arms outstretched and heads tilted. The circle and the people in it start slowly and go faster and faster for as long as twenty minutes—until finally they stop, almost abruptly. You expect them all to fall down in a pile, like five-year-olds, but they don't. The whirling is a form of meditation; in the ecstatic state, practitioners describe a feeling of union with god. "Do you know what the whirling is?" wrote the thirteenth-century Sufi mystic and poet Rumi. "It is . . . opening the eyes of the heart and seeing the sacred lights."

In modern times, artists and writers have not given up trying to describe the moment when a soul beholds God. No one has done it better than C. S. Lewis in *The Lion, the Witch and the Wardrobe,* in a scene where the four children finally see Aslan, the lion king (and Christ figure) they had heard so much about: "People who have not been in Narnia sometimes think that a thing cannot be good and terrible at the same time. If the children had ever thought so, they were cured of it now. For when they tried to look at Aslan's face they just caught a glimpse of the golden mane and the great, royal, solemn, overwhelming eyes; and then they found they could not look at him and went all trembly." The lion knows who the children are without being told, of course, and when they come to stand before him, "his voice was deep and rich and somehow took the fidgets out of them. They now felt glad and quiet and it didn't seem awkward to them to stand and say nothing." I read the Narnia books when I was a child without knowing they were Christian parables, and I can still remember how I felt upon finally meeting Aslan. He was magnificent and brave, familiar, warm, and comforting.

A face-to-face encounter with all-powerful God is not always cozy, and Bee Season, published in 2000, offers another heavenly meeting, this one as terrifying as a horror show. The heroine of Myla Goldberg's novel is Eliza Naumann, a young girl with a dysfunctional family and a savant-like gift for winning spelling bees. She becomes enamored of and then obsessed with Kabbalah mysticism, the repetitive chanting of certain symbolic letters and verses in the Zohar. One night, alone in her room, while her mother is locked away in a mental institution and her father is arguing with her older brother, who's become a Hare Krishna, Eliza ascends through the letters and words to God himself. "She is mauled and licked clean by a lion. . . . She is choked and crooned to, slashed and kissed, stung and cuddled, fed upon and fed. . . . This is the pain of creation, of life emerging from void, of vacuum birthing being." This picture, like those angels in Revelation, describes a heaven not wholly pretty, and never placid.

WE WILL SING

Not only do souls see God in heaven, they praise him, constantly, with song. To their arsenal of heaven's pleasures, the faithful always add singing. The Torah and other Jewish scriptures are full of music. Miriam, the sister of Moses, has a tambourine. David, the king of the Jews, has a harp. David's poems, the psalms, are full of exhortations: sing praises to the Lord and blow trumpets in joyous celebration of him. For the Jews who kept the great library at Qumran, which came to be known as the Dead Sea Scrolls, the connection between singing on earth and singing in heaven is clear. One Qumran author writes that while he was singing prayers of thanks to God, he was raised "to the height of the universe . . . to stand among the ranks of the host of the holy ones and to join together in the assembly of the sons of heaven." These first-century Jews understood singing as something angels did in heaven—and as something people did on earth that made them more like angels.

The book of Revelation—the heavenly Christian vision with Jewish roots—is also full of singing. There are those terrifying angels, buzzing around the throne of God and singing "Holy, Holy, Holy." Twenty-four elders with harps stand before the throne: they're singing, too. Surrounding the throne and the elders are "ten thousand times ten thousand" angels, and they're also singing. Music reverberates everywhere, around "every creature in heaven and on earth and under the earth and in the sea, and all that is in them." And this is the song the creatures in heaven sing: "To the one seated on the throne and to the Lamb be blessing and honor and glory and might forever and ever!"

The earliest Christians understood that when they were singing God's praises in unison they were participating in the activities of heaven. In the fourth century, the Syrian preacher John Chrysostom made the connection explicit. Don't worry about the quality of your voices, he told his listeners, don't worry if you're old or young. Just sing. "Above, the hosts of angels sing praise; below, men form choirs in the churches and imitate them. . . . The inhabitants of heaven and earth are brought together in a common solemn assembly; there is one thanksgiving, one shout of delight, one joyful chorus." The stained glass windows in many of Europe's great cathedrals show angels singing. They serve as a constant reminder: when a choir on earth sings, the angels do, too. In a 1995 sermon, Cardinal Joseph Ratzinger—who was to become Pope Benedict XVI—declared that "the sacred liturgy is not something which the monks manufacture or produce. It exists before they were there; it is an entering into heavenly liturgy which was already taking place." The singing in heaven is happening all the time, and humans can join.

In an essay on heaven, the novelist Rick Moody describes the fearsome organ music he heard in church as a child. Bach, he says, echoing many before him, will certainly be *Capellmeister* of the heavenly choir. ("It may be," wrote the philosopher Karl Barth in a 1956 letter on the two hundredth anniversary of Mozart's birth, "that when the angels go about their task of praising God, they play only Bach. I am sure, however, that when they are together en famille they play Mozart and that then too our dear Lord listens with special pleasure.") And though Moody knows that choral singing is traditionally the musical activity in heaven, he can't resist putting a contemporary, individualistic spin on the concept. A musical omnivore, Moody hopes that in heaven God will be in possession of some kind of massive figurative iPod, which plays everything he wants to hear exactly when he wants to hear it—even songs with profane lyrics, even songs by cheesy bands like Rush.

Not everyone imagines the singing in heaven to be harmonious, however. In most Puritan churches, harmony singing was strictly forbidden. Even today, in the churches of the Old Regular Baptists, the congregation sings without accompaniment and with some members "lining out," or "curving"—intentionally fraying the unity of the song. The idea, born around the time of the Reformation, is to keep congregants focused on the words, not the tune, and to help everyone—even people who couldn't read from the hymnal—to join the chorus.

My family and I belong to a Reform Jewish temple in Park Slope,

Brooklyn. As I became more immersed in this project, I began going more frequently to temple and I found that the service I liked best was the family service, at nine-thirty on Saturday mornings. It's a very informal, coffee-and-bagels affair. Parents come in jeans, and there's a rug so the littlest kids can sit on the floor. Although some members of my family are extremely musical, I am not, and one of my human failings is my relative lack of interest in music (despite my love for David Byrne's *Heaven*). However, the family service has a musical ritual that brings me to tears every week, without fail. The most ancient Jewish prayer is the Shema. It's at least twenty-five hundred years old and staggering in its simplicity: "Hear, people of Israel, the Lord is our God, the Lord is One." It's a declaration of belief in One God, and when you realize how fragile that belief was all those years ago, and how miraculous it is that this particular prayer—and the Jews themselves—have survived, and when you think about all the generations of Jews who have intoned these same words for so many centuries all over the world—in Poland, Belgium, Holland, Austria, Lithuania, New York, Chicago, Los Angeles, and Ohio in my family alone—it's enough to make anyone weep.

At the family service, though, the rabbis in their wisdom have doubled down on the poignancy. The Shema is sung by the congregation's kindergarteners and first-graders, and to accompany the singing, the rabbis have taught the kids the words in the international sign language for the deaf. So there they are, with their small faces, uncombed hair, and Ugg boots, singing the ancient words and signing for the deaf, and I'm a basket case. In those moments, I am aware, all at once, of the ancient history of the Jews, of the miracle of human life and human love and its fragility and brevity; and I sing, and my daughter sings, too, and we are united with one another and with all of the Jews who came before us and all of those who will come after.

Ancient ideas of heaven come alive in those moments. We are praising God, together, in a community, in history, out of history, in the context of overpowering love.

Paul Westermeyer is a historian of church music at Luther Seminary in St. Paul, Minnesota, and he believes that the feeling I have during the Shema—and that so many people describe having in worship—is the best answer to the question of whether heaven is boring. I spoke to him at length one winter weekend afternoon, just after he'd finished installing new carpeting in his family room. "It's hard to imagine anything less boring than seeing God and singing to him," he says, "if God is worth anything." Westermeyer believes that when people sing together the angels sing with them—and that when people stop singing, then heaven grows silent.

The problem with too much American worship today, Westermeyer says, is that it's conceived as entertainment, not as something participatory and transcendent. Too many Christian pastors concern themselves with growing their churches, Westermeyer says, and they see music as a marketing tool, a way to pull in the greatest number of desirable people. In these churches people sit and listen to a band during the service; they sing, karaoke style, as the words to the hymns appear on a giant movie screen that hangs behind the altar. They are passive, not active; they are earthly, not angels. But Westermeyer has hope: all across the country are small churches where singing together—whether it's folk or gospel or traditional hymns—is alive and well, and in those churches, he says, the people are, in some small way, experiencing heaven.

Heaven is certainly alive and well at Transforming Life Ministries, a small Pentecostal congregation that used to worship on the second floor of a falling-down building in the Crown Heights section of Brooklyn. (It has since moved to another part of the same neighborhood. Half a block away, across Eastern Parkway, the residents are all Orthodox Jews.) One Christmas, I went there with Beverly Adjodha, who has taken care of our daughter since she was four months old. Beverly came to Brooklyn from St. Lucia in 2001, when she was twenty-four. Though raised a Roman Catholic, Beverly was, immediately upon her arrival, adopted by this small, vibrant congregation to which her aunt belonged. Soon it was the center of her life. Beverly goes to church three nights a week and all day on Sundays. She met her husband there, and when she became pregnant, the church threw her a baby shower with gift bags for everybody. Pentecostalism, the fastest-growing brand of Christianity in the world, is characterized by a belief in "the gifts of the Holy Spirit"—outward manifestations of God that include healing and speaking in tongues. There are over 550 million members of Pentecostal and related charismatic movements in the world today. Pentecostalism is growing so fast in the developing world that the Boston-based minister Gene Rivers predicts that within a generation, it will become the dominant form of Christianity everywhere.

As I ascend the dark stairwell to the church, I can't believe I'm in the right place. The hallway is unlit, the stairs are rotten. Then I push open the door and face a brightly lit maze of rooms: women in suits and dresses are preparing breakfast as children learn Sunday school lessons. At ten a.m., the singing begins. Most Christian services warm up with two or three hymns, but at Transforming Life, the initial singing goes on for a full hour.

Up on the dais—no pulpit here, only a lectern in front of gauzy gold curtains—there's a small band: two electric guitarists, a bass player, and a drummer. The congregation is comprised mostly of West Indian immigrants, and the music has a calypso lilt, even though today's songs are such conventional Christmas carols as "Oh, Come All Ye Faithful" and "Good King Wenceslas." There's a song leader, a middle-aged woman whose brow furrows intensely in concentration. Four women flank her, two on each side. None is dressed alike. All five sing with their eyes closed and their palms upward. I have never heard anything like this. The singing is strong—jubilant—but not exactly harmonious, and always, in the background, is the babble of people speaking in tongues.

During the singing, the people rise from their folding chairs and dance. Across the aisle, a woman tilts her chin up as she sings, sways her hips, and makes big, cupping motions with her arms, as though she's trying to pull the Spirit nearer to her. In the front row, the minister's wife has both arms in the air and hops from one foot to another in a jagged little dance. As the hymn fades away, the babble in the background grows louder. Up on stage, one chorister is murmuring, "Praise God, praise God, praise God, praise God," while another is saying, "Hallelujah, Hallelujah, Hallelujah, Hallelujah." A third intones, "Oh most excellent king." The babble, dissonant, cacophonous—lasting sometimes for as much as ten minutes—reminds me of those angels in Revelation singing "Holy, holy, holy." When I ask the pastor, Albert Paul, what he believes people do in heaven, he says, "We will do the things of God. We will rejoice. We will be with Him, from morning till night."

WE WILL LEARN

Still, David Byrne has a point. All of this seeing God and praising God may be wondrous, but it's static. It doesn't "get" you anywhere. Christian theologians might argue that in heaven, ambition, competitiveness, drive—and time itself, in which such notions make sense—cease to exist. But for the kind of person who doesn't like to read at the beach, the absence of forward motion sounds like a jail term. For them, a world in which achievement doesn't matter can't mean very much. Going to heaven has to mean *getting on with it*. There has to be a notion of progress.

Earlier Christians had no idea of progress in heaven. In Dante's conception, heaven has nine levels but every saint is happy within his or her designated place. By the nineteenth century, though, "progress" had become a Western ideal, and a number of Protestants began to conceive of a heaven in which a person could continually improve, ascending higher and higher, attaining greater and greater understanding of God—who, in His infinitude, gives them more and more to learn. In Swedenborg's vision, the more a person loves God in heaven, the more that individual becomes fully himself. Striving persists in Swedenborg's heaven, though the angels get occasional days off for relaxation and pure enjoyment. Latter-day Saints refer often to postmortem progress, to the idea that the afterlife is a continuation of this life. "There's so much work to be done [in heaven]," said the religion's founder, Joseph Smith, "we need to be in a hurry in order to accomplish it." After death and the resurrection, wrote Methodist minister Jeremiah Dodsworth in 1853, the saints would be busy attaining "higher and still higher degrees of perfection." When I went to Tampa in 2007 to interview a group of Vietnam veterans about heaven, I was especially struck by what their pastor—a Methodist chaplain with strawberry blond hair and a wide-open face—said he believed. "Well," he said, as we sat eating cheeseburgers at a Perkins Family Restaurant, "we have to work in heaven. I think God made us to be people who have to work and strive and that's how we find happiness. . . . I can only stand so much vacation."

In the nineteenth century, a group of Muslims calling themselves the Ahmadiyya movement broke away from mainstream Islam to follow an

Indian Messiah named Mirza Ghulam Ahmad. The Ahmadis—who still consider themselves Muslims, and claim tens of millions of followers worldwide—believe strongly in the idea of progressive enlightenment in heaven. On Judgment Day, wrote Ahmad in his 1896 essay "The Philosophy of the Teachings of Islam," the righteous will be offered the gift of perfect light, and "their supplication for the perfection of their light indicates that their progress will be unlimited. They will attain one perfection of light and then will behold another." Imagine Dante's tiers of heavenly light combined with forward movement.

Within Jewish thought, this strain is stronger still. Almost every Jew I spoke to joked about heaven being a place where the rabbis are sitting around learning Torah. Rabbinic authorities even speak of scholars who die as having been summoned "to the Academy on High," where departed sages pass eternity studying and debating theological minutiae with God himself. "Heaven is a Talmud class, without recess and without lunch," quipped David Berger, the noted Maimonides scholar, recalling a joke a classmate made in college. This idea, of heaven as a place of continual, endless study, is based on part of the Talmud: "In the world to come there is no eating, or drinking nor procreation or commerce, nor jealousy, or enmity or rivalry—but the righteous sit with crowns on their head and enjoy the radiance of the Shekhina [divine presence]." The rabbis, who considered the ultimate spiritual activity to be learning Torah, interpreted this passage as a Torah class and imagined they would enjoy the divine presence in this fashion. Berger explains the movement of souls after death this way: "People speak about rising from one level to the next, because our soul is learning things. This is a class. But since God is infinite, there is an infinite amount to learn."

When Bob Dylan decided to study Judaism in 1982, he called upon the Lubavitch rabbi Manis Friedman. The Lubavitchers are a group descended from Jews who lived in Russian shtetls in the nineteenth century. Their approach to Judaism emphasizes spiritual and mystical experience, not dogma. Friedman is a scholar, a television host, and the author of *Doesn't Anyone Blush Anymore?: Reclaiming Modesty, Intimacy, and Sexuality.* I met him one day in Rockefeller Center. He was easy to spot: a handsome, white-bearded rabbi with warm brown eyes behind wire-rimmed glasses. We talked, as planned, in the back-seat of a banged-up car as a young Israeli drove Friedman to La Guardia Airport, where he was to catch a plane home to St. Paul, Minnesota. Friedman speaks of heaven as a place where souls are active, where they continue to ascend, to learn, to perfect themselves. Souls, he explains, are eternally alive. They cannot die. "What is dead can't live, what is alive can't die. Then what does death mean? This living soul, which cannot die, enters a body, which cannot live. For 80, 90, 120 years, the body lives off the soul. When they separate, the soul continues to live. The soul goes back to the universe of souls. We call it heaven."

"There are different places in heaven," Friedman continues. "Endless levels. Since God is infinite, there is infinite closeness that God can achieve. Heaven is not stagnant. You're always growing, you're always moving. In heaven, they don't rest. The body should rest in peace. The soul does not rest. When a soul reaches a really high level of closeness to God, it becomes more and more like God and then it wants to come back to earth." Is Rabbi Friedman talking about reincarnation? "Yes."

At least since the Middle Ages, reincarnation has been a strong thread in Jewish thought. Discounted by certain rabbinic authorities, including the tenth-century philosopher Saadiah Gaon—and regarded as fringe by most contemporary Conservative and Reform rabbis—reincarnation is central to Kabbalah mysticism, and thus plays a role in the life of any Jew who regards himself as an heir to the mystic tradition. Kabbalah has its roots in ancient oral traditions, but became more widely known in the twelfth century, when Spanish Jews began to publish tracts outlining the movement's doctrines and symbols. The most important of these texts was the Zohar, a series of biblical commentaries written in stilted Aramaic in the thirteenth century and traditionally attributed to the second-century Palestinian teacher Simeon bar Yohai; kabbalistic mystics meditate on the text, and on individual letters within the text, and in so doing produce "out of body" experiences. Kabbalists speak of *gilgul*, a revolution of souls, in which souls ascend to God and then back down to earth on a sort of invisible Ferris wheel. Those souls that have attained perfect righteousness remain on high; those in need of further perfecting—either to fulfill more mitzvoth or to cleanse themselves of sin—return for another try. "Truly," says the Zohar, "all souls must undergo transmigration; but men do not perceive the ways of the Holy One . . . not the many transmigrations and the many mysterious works with many naked souls." Today, most Jewish adherents of reincarnation tend to be among the Orthodox, explains David Berger of Yeshiva University—for the simple reason that the Zohar and other mystical texts have been "accepted either overtly or tacitly by most modern rabbinic authorities."

The prevalence of mystical belief among Jews has had a viral effect. The Kabbalah Center, with headquarters in Los Angeles, dispenses Jewish mysticism to the masses—though serious scholars scoff at the simplistic version on offer there. It teaches reincarnation, though, which may explain the center's popularity with celebrities: Why reside forever in heaven when life on earth is so good? As noted in chapter 5, nearly 30 percent of Americans say they believe in reincarnation. "We all want the here and now, and reincarnation is about the here and now," Steve Prothero told me in an e-mail. "Reincarnation is fueled in part because now people WANT to come back and live again. Of course, the whole idea before was to get OUT of the cycle and not to stay in it. Now we WANT another life." Before, life was hard. Now, at least in Western countries, life is bountiful, blessed, and fun.

WE WILL BE ONE

Early in the second act of *West Side Story* is "Somewhere," Leonard Bernstein and Stephen Sondheim's vision of heaven. In the 2009 Broadway revival, dancers from the two warring tribes meet before a shimmering silver-blue scrim. In pairs that real life would make impossible, the couples dance sweetly, chastely, delicately. It's a beautiful dream in which the only thing that matters is love. As every fan of *West Side Story* knows, the marriage of Tony and Maria can only endure somewhere else, in another time, another place, after death, in heaven.

This vision—heaven as a place for all followers of the same Truth—is a progressive interpretation of an orthodox view. You can't get to heaven through belief in Christ or the Qur'an, in other words. You get there through your commitment to Love. Or Justice. That, anyway, is what Salman Ahmad believes. Ahmad's beautiful face could only belong to a rock star—and it does. He is the lead singer and guitarist in a band called Junoon ("obsessive passion" in Urdu)—one of the most popular bands in Pakistan. The *New York Times* has named Junoon "the U2 of Pakistan," and called Ahmad "a figure like Bono." For Ahmad, heaven is the abode of the people of God in a global world.

Ahmad was born in Lahore, Pakistan, to parents who dreamed the American dream. (His mother had spent some time in California as a girl and always yearned to come back.) His father, an airline executive, moved the family to the small colonial town of Tappan, New York, when Salman was eleven, and the boy seamlessly entered the life of a suburban teenager in the 1970s. He listened to Led Zeppelin and smoked marijuana. He taught himself to play guitar and joined a band. His priority was then—as it is now—to emphasize the sameness he felt with his white Jewish and Irish-Catholic friends. Not difference. "I don't want to belong to any tribe," he told me.

From his earliest childhood—even when he still lived in Pakistan—Ahmad rejected organized religion. His grandfather on his father's side was very pious. He prayed over every mouthful of food—and often begged for young Salman's company during Friday prayers. Ahmad always found an excuse to say no. Too much homework, he would say. Going to the cricket match. "For me," he remembers, "that was, like, man, you're turning us into robots." Later, as Ahmad entered high school in Tappan, his mother began to worry about her oldest son's soul. Never very observant herself, she began to nag her boy the way traditional Muslim mothers do about his spiritual seriousness and direction in life. She would urge him to look at the floor respectfully when he passed a girl in the hall at school. She pressured him to become a doctor. She agonized over whether her Muslim son was becoming too American. "I would have none of it," Ahmad says. "I just rebelled. This was 1970s America in the age of sex, drugs, and rock and roll, and the pressure from my family was driving me to be a schizophrenic. I chose to be a rocker." His mother had the last word, though. When Ahmad was eighteen, the entire family moved back to Pakistan.

My family and I piled into the car one Saturday and drove to Tappan, where Ahmad lives once again, now with his wife, Samina, and their three teenage sons. They are Americans. Ahmad speaks about waking up before dawn on Election Day in 2008, so excited was he to go vote for Barack Obama. The Ahmads live in a small, two-story house marked by large wagon wheels in the driveway. Inside, the living room is furnished with low couches slung with gray and blue silk pillows. Above a sofa hangs a framed Punjabi inscription that Salman translates as: "God lives in your heart. He is not unreachable." We sat, eating chips and Samina's famous biryani—rice with vegetables and chicken—until day turned into night. Salman and I drank red wine. For dessert, there was apple pie and ice cream.

Ahmad says he became a student of Islam after falling in love with Samina, whom he met at a wedding in Lahore. "Within this chaos, there was beauty," he says. "Within this chaos there was love." She was devout. Thanks to her, he began reading the Qur'an, and also the Sufi mystics—especially the poet Rumi. He became a doctor to please his mother. He started a rock and roll band to please himself. Junoon sounds like what it is—South Asian psychedelic rock, both hard driving and euphoric. The band plays wherever Muslims live, even—or should I say especially—in places like Peshawar, where live rock concerts are neither normal nor welcome. Ahmad sets Islamic poetry to music. The lyrics to the song "Khudi" ("Self") are based on a poem by the twentieth-century Pakistani poet and philosopher Muhammad Iqbal:

There are worlds beyond the stars
Many trials of faith and desire still to face
You are the falcon
It is in your nature to fly and more horizons for you to transcend
Raise yourself so high

I imagined Ahmad would tell me that heaven for him was in the transcendent feeling he got from playing music. He did not. Ahmad believes—both ideologically and spiritually, if you will—in Oneness. He believes, as so many have said, that God does not care about religion: "Ritual, dogma—it's man-made." For Ahmad, heaven is in the perfection of people doing something sublime together, so the whole is more than the sum of its parts. Like what? He gives a surprising answer. "The Pakistani national cricket team, in 1992. They came back from the dead to beat England that year. They were so behind, one Pakistani newspaper wrote 'Bury Them.' To understand each player's unique personality and gifts, the harmony between free will and predestination, how it worked, it was like . . . wow."

Ahmad believes—some would say naively—that heaven consists of the Sunni and the Shia and the Buddhists and the Hindus and the Jews and the Muslims and the Christians getting along. He sees heaven, as he did in 2008 when he performed at a benefit concert for the Muslim Public Affairs Council in Los Angeles, in the ability of his friends Rick Warren, who opposes gay marriage, and Melissa Etheridge, who is gay, to see the best in each other, and he refutes my skepticism on this matter by looking at me, silently smiling, and drinking his wine. He calls this central idea of his *tawhid*, "divine unity" in Arabic. It is the driving force in his life. You'll notice that he answered my question—"What is heaven?"—by not really answering it at all.

EPILOGUE

I was with my grandfather when he died. It now seems like something that happened in another lifetime, and in a lot of ways, it was: before Charlie, before Josephine, before our mortgage and the crushing busyness of our lives—also before the extreme joy I feel at seven-thirty each morning, when Josephine comes into our bed for a snuggle and, for ten or fifteen minutes, I try to make time stand still. *This is everything*, I tell myself. *This is all there is.*

In that other lifetime, I was a runner. It was August, and I had just come back from a long Saturday run. I was standing in my bedroom, dripping with sweat, when the phone rang. "Please come," my grandmother said from their apartment uptown. "Your grandfather is dying."

This was not a surprise. My grandfather had been terminally ill with cancer for six months, and we had been through all the usual ups and downs. Hospital stays that turned from days to weeks, bitter fights with doctors we considered unsympathetic or incompetent, dehydration, rehydration, deathbed reconciliations, dramatic recoveries, delusional nights, vivid and articulate mornings. By the summer he was at home, existing on Ensure, a sweet, fatty liquid like baby formula that nourishes so many people at the ends of their lives. We were lucky to be able to hire nurses to care for him 24/7—angels, my mother called them. And they were.

Through his illness, my grandfather had remained astonishingly lucid. He had always had a gift for languages, and the circumstances of his life—born in Belgium, proprietor of a business that took him all over Europe and Eastern Europe, an immigrant to America—gave him ample opportunity to practice. He spoke—fluently—French, English, German, Flemish, and Yiddish, and—less well—Spanish, Italian, and Polish. During his hospital stays, he flirted with the nurses in whatever language they spoke. He could do the *New York Times* crossword puzzle in ten minutes on any day of the week. My mother says he was a beautiful figure skater, though I don't believe I ever saw him skate. He could whistle with a trill, and most often he whistled Mozart. He was, in other words, a cultured, formal, European man from a bygone age. He dismissed all talk about God or organized religion as silliness—and even teased me about my job as religion writer. Yet it was his Jewishness that had forced him to pack the car that day in May 1940, leaving everything he knew behind.

I had seen him just a few days before my grandmother's call. He was always happy to see me. His eyes would brighten, and he'd hold my hand. His hand, which had a tremor for as long as I can remember, would be dry and warm. We'd talked that day about his childhood in Antwerp: about his family's mustachioed chauffeur, Charles, who taught him how to drive; about the time he kicked an apple all the way to school until it was mush; about the time he picked up a stick in a field and it turned out to be a snake. We played backgammon, and he beat me.

When I arrived at the apartment that morning, my grandfather was lying in his bed with his eyes closed. He skin was waxy. He was breathing gently: his exhalations went "puff puff"—not labored, but as if he was running out of gas. Circling his bed were my grandmother, my great-aunt Anita (my grandfather's little sister, always spoken of as the great beauty of the family), and the weekend nurse, Michelle. We all just stood there, silently. I touched my grandfather's abundant hair, which he had always worn neatly slicked back with oil. It felt silky.

The unforgettable thing about his death is its gentleness. There was no before-and-after moment, no final cry, no eyes wide open, no last words. None of us standing there, frankly, could tell when it happened. He made a "puff puff" slightly louder than the ones before. Perhaps that was the moment. Finally, Michelle found a mirror and put it to his mouth. No fog. I touched his forehead and it was cool. That was it. My grandmother wailed, "Where is my family?" I called the funeral home.

I tell this story because in the course of writing this book, whenever I have asked myself—over and over—"Do you believe in heaven?" I always think of my grandfather. I try to visualize him. I loved him, I was there when he died; I miss him and my grandmother every day of my life. Surely, if I believed in heaven, I would see them there in my mind's eye.

Sadly, I don't. When I ask myself, "Where is he now?" all I see is the cemetery in Westchester, the shady hillside where both he and my grandmother were buried—he on a sweltering day, she in chill January rain. I do not envision my grandparents alive anywhere. I did not see, or even imagine, my grandfather's spirit rising from his body that morning, and I have never felt him looking down on me. (Although I must confess that when I use the blue stew pot my grandmother gave me from her own kitchen cabinet—something I do three or four times a week—I can see her freckled hands on it, holding the rim with a battered pot holder and stirring the pot.) I do not believe in a supernatural realm where my grandparents exist as themselves, nor do I imagine them engaged in any of the activities they loved on earth: watching tennis on television, say, or sucking on hard candy and listening to chamber music at Lincoln Center. Although I do believe the world will end—everything ends—I do not believe that end will be accompanied by glorious resurrections.

Nevertheless, my experience of my grandparents' deaths links me to my ancestors and my tradition in important ways. My grandfather was eighty-nine when he died. He had his family around him and he was buried among his people—or at least, among other New York–area Jews, who might well have been his people. His children were prosperous and had children of their own. He had heroically preserved his own lineage when he fled the Nazis. He was—as anyone in my family will tell you—our clan's patriarch. Whenever any of his children, grandchildren, nephews, nieces, in-laws, or cousins needed anything, particularly advice or money, he would always oblige. He may not have believed in heaven, but he believed in family. He wouldn't have put it this way, but he believed that the accomplishments and honor of one generation accrued to the next, and that it was the responsibility of children to safeguard and perpetuate the legacy of their parents. He was our Abraham. He "died in a good old age, an old man and full of years, and was gathered to his people."

Also, I had a little visit from him after he died. I was a skittish bride, terrified that my marriage would somehow be a great mistake. My grandfather had died about six months before I met Charlie, and though I was sure the two of them would have liked each other, the missed connection bothered me. (Also, to state the obvious, Charlie is not Jewish and I wanted my grandfather to tell me that was okay.) Then, in the middle of the night a few weeks before my wedding, I woke to see my grandfather sitting at the foot of my bed. He looked well, the way he did before he was sick, his hair slicked back and his saggy brown eyes twinkling. He said nothing, but he patted my arm. In the morning I felt reassured, as if I'd gotten his blessing. He may not have been real, but he felt real to me, and if anyone had suggested the next morning that I pay tribute to his spirit in some way (as my Hebrew ancestors did), I would have. I did not belong to the temple at the time of my grandfather's—or grandmother's—death so I did not say Kaddish. But I do now, in the season of each *Yahrzeit* (death anniversary), bumbling along in Aramaic.

My biggest beef with the atheist writers Christopher Hitchens and Sam Harris—whose works I generally admire—is this: the God they set up to knock down is a straw man. They define God in the most simplistic, kindergarten terms: an omniscient, omnipotent meddler, whose inscrutable ways and insane approach to justice motivate people to great evil, or, at best, to compliant thoughtlessness. (Harris famously compares the God of Abraham to Zeus.) But while many people do believe in such a God—and, indeed, perform ruthless acts of murder and intolerance in His name—others don't. Just a third of Americans believe in a God who controls human events, according to a 2006 Harris poll.

It may be naive to say this, but "God" is the word I use to describe what is miraculous about this life, the aspect that is awesome and defies rational explanation. Water equals two hydrogen molecules plus one oxygen molecule, yes. But . . . you can drink it! Fish can live in it! It fills our bodies and covers the earth! Josephine equals one part Charlie and one part me plus—a very large part—something that is just Josephine, seemingly begotten out of thin air, unlike either one of us and so very much like herself. Okay, I'll stop. But you know what I mean.

The God of which I speak, then, is, in some sense, a creator. Not a guy who put the whole thing together like Tinkertoys in six days, but the miraculous aspect of the process of creation, the fact that the earth, the water, the animals, and the plants work together to support human life and that from human life comes creativity and achievement, and even a sense of the Divine. I'm not speaking literally here. I believe we can break down all of God's creation into molecules, name them, sort them, catalog them, and discover their origins, eventually. All I'm saying is that there's a "holy shit" aspect to all of it that's worth contemplating from time to time—and when I do contemplate the places that bear the stamp of God's creation (or inspired human achievement)—whether the great cathedrals of Europe or the land-scape of southeastern Utah—I find heaven. The nineteenth-century Dutch painter Vincent van Gogh had a continuing, dissatisfying re-lationship with the Protestant church of his childhood, of which his father was a minister. Nevertheless, he wrote in a letter to his brother Theo, this did not keep him "from having a terrible need of—shall I say the word—religion. Then I go out at night and paint the stars." I need not add that van Gogh's *Starry Night* is one of the most super-naturally vivid renderings of heaven ever created.

Progressives like to say that God is love, and while Hitchens and Harris might sniff at that, it's a notion that I can get behind. If God is love, and heaven is where God lives, then heaven exists in the love between people—and between people and God.

I think all the time about a woman named Agnes Long. In the early phases of researching this book, I became obsessed with the Desert Fa-thers, the monks and hermits who took themselves out of society and into the deserts of the Middle East in the first centuries of Christian-ity, where they lived in extremis—with little food and water, little conversation or companionship, just the austerity of the landscape and constant prayer and contemplation. I discovered, through my research, that a few Roman Catholic dioceses were permitting lay-people—people who had formerly had regular jobs, spouses, and families—to live as hermits under their auspices. These people would generally give away all but their most necessary belongings, and retreat to the woods or the country somewhere, to live out their lives contemplating God. I felt I needed to meet one of them, and after months of searching, I did. Agnes Long lived on Wisconsin's Madeline Island, in the middle of Lake Superior. To get there from New York you have to fly to Duluth, Minnesota, drive two hours, and then, if it's winter—as it was when I visited—take an ice shuttle across the lake. I wrote a story about her for *Newsweek*.

Evelyn Agnes Sebastian grew up a sad and lonely child in New Jersey in the years before the Second Vatican Council. Her favorite pretend game was "nun." She made herself a little chapel in her bedroom and designed and sewed her own habits. But as she got older, it turned out she was gorgeous. Blond and curvy, she worked as a model for Sears, and before long she was married and the mother of three children. Hers was not a happy marriage, she told me, and after nearly twenty years, she got divorced. She entered a party-girl phase—she had a well-paying job, she bought expensive clothes, she went to clubs in Manhattan. It was during that phase that she met Marvin Long.

"He was the love of my life," she tells me, and from the pages of her Bible she pulls an old color photograph. They are standing together on a boardwalk at the Jersey shore. He is tall, dark, and fit and wears a bathing suit. She wears a floppy hat and a zebra striped bikini. Together they moved to Texas—he was in the oil business—where they had a comfortable life: a big house, three cars, and a bedroom set made entirely of glass. In 1984, Marvin died of a brain tumor and Agnes sold everything and began the long, slow process of giving her life to God. In 1997, she took vows of poverty, chastity, obedience, and solitude, and moved into her hermitage (a tiny, prefab wooden house) on Madeline Island. When I met her, she had a phone but she rarely used it. She almost never saw her children. She spent her days painting religious icons and praying the Divine Office—regular cycles of prayers, five times a day. She ate one meal a day; her food was given to her mostly by kind neighbors. She wore a denim habit—blue, she told me, because it represents the Virgin Mary and denim because it represents the common man.

Here's the thing about Agnes Long. She disappointed me. I had traveled all this way to meet one of the Desert Fathers (or in this case, Mothers)—a wise, older person who, I hoped, would share with me her profound experience of encountering God—and heaven—all the time. Instead I met a heartbroken woman, who lived in a depressing shack in the middle of the woods and was entirely unable to articulate what had moved her to make this obviously radical decision about her life. Where is heaven? I kept hounding her. What do you think about when you think about heaven? What do you read to think about heaven? Finally, she grew irritated with me. "You talk as if heaven is a *place*," she said. "What I think of is the Lord. At different times I have felt his presence in that joy that has no reason or explanation. It comes and goes. Heaven is being totally united with God in love."

Ah. This is something that, years later, I have come to understand in my own life, with Charlie and Josephine—our continuing struggle to be kind and respectful to one another in spite of our human selfishness. One of Christianity's most resonant metaphors for heaven is marriage, and though I am not a Christian, this metaphor makes sense to me. In our communities, in our families, we love one another all the time as best we can, fully conscious of how difficult it is and how much humility is required. Josephine had a book when she was littler called *I Love You All the Time*, in which a mommy bear tells her baby that she loves him all the time, even when she's with friends, or at work, or on the telephone, or busy with something else, or when she's angry with him. "Even when you can't see me," she concludes as she turns off the light in his bedroom, "I love you all the time."

Heaven is in there, somewhere. The joy of that love is—truly—beyond anything I ever imagined.

I am a progressive in my heart, but I yearn at times for the discipline and the faith of the orthodox. I wish I could somehow "go there" and embrace the supernatural aspects of heaven—the streets of gold, the many mansions, the banquet, the Torah study, the music, the physical enjoyment of all kinds of pleasures, the bliss, the reunions. What a comfort it would be to me—and even more to my skeptical friends who have lost children. I even yearn for that literal-plus interpretation of scriptural descriptions given to me by believers who are also intellectuals. The heaven of the Qur'an "is described to us by names we can understand, but it's a completely different experience," says Hisham Abdallah, the Roche pharmacologist who lives in Silicon Valley. In other words, Abdallah believes that in paradise he'll see green, green pastures as the Qur'an promises, but he also believes that those green, green pastures mean something else entirely. I wish I felt that.

What I don't wish for—this is my line in the sand—is any certainty about who's in and who's out. Any sense of hierarchy among good people—one Mormon sister's in because she obeys the rules, another's out because she doesn't—infuriates me, as does any Christian, Muslim, or Jew who believes that there are only Christians, Muslims, or Jews in heaven. I said as much in chapter 6: I have nothing but admiration for the evangelist Billy Graham, who had preached for his whole life that belief in Jesus Christ was the only way to heaven and then, in his last decade, softened his stance. "I think [the Lord] loves everybody regardless of what label they have," he told my boss, Jon Meacham, in an interview that appeared in *Newsweek*. I hope Graham's children, some of whom are more conservative on this matter than he was, took note.

When conservative Christian, Muslims, and Jews talk about heaven, they often use the word *radical* to describe what they mean.

The heaven that will come at the end of the world is a *radical* reversal of the social and natural order. The first shall be last, the meek shall inherit the earth, the stars will fall from the sky. Heaven, for them, is not just love, it's radical love; it's not just a return to the perfection of Eden, but a radical return. This is no warm hug, no easy train ride. It's radical because God is involved and God can do anything. While I do not believe in this intervening God, I do cling to the idea of heaven as a radical concept, a place that embodies the best of everything—but beyond the best. A belief in heaven focuses our minds on the radical nature of what's most beautiful, most loving, most just, and most true. At the beginning of this book, I said I believed that heaven was hope. I would now amend that to say, "radical hope"—a constant hope for unimaginable perfection even as we fail to achieve it. As Emily Dickinson said, heaven is what we cannot reach. But it is worth a human life to try.

AUTHOR'S NOTE AND ACKNOWLEDGMENTS

This book is—as are all books, I suspect—a personal journey. It's the story of one woman exploring more than two millennia of Western ideas about heaven. As such, it is not comprehensive. Readers may complain that I have not included their favorite heaven joke, painting, song, or literary reference. Some will certainly observe that John Milton and *Paradise Lost* receive only passing mention, while thousands of words are devoted to its predecessor Dante's *Paradiso*. They will have observed rightly. Heaven is infinite, as are thoughts about and interpretations of heaven. While I do not aim to be inclusive, I did try to write a book that's broad and balanced enough to give every interested reader something to chew on. Its idiosyncrasies are fully mine.

I have been astonished at the generosity of the scores of scholars, clerics, and laypeople who became characters in this story. Not only did they give me their time—repeatedly—but many spoke to me about their deeply held religious beliefs, as well as about personal tragedies and inner conflicts. These people are the soul of this book: their thoughts and visions enliven and contradict historical facts and theological abstractions. Faith is a complicated business. Thank you to all of those who articulated those complications so eloquently.

Over six years, I came to assemble what I call my panel of experts. These are scholars in the field of religion who have a particular interest or specialty in heaven and/or American religion. They have been—without exception—unfailingly generous with their insights, answering every kind of question by e-mail or phone at odd hours of the day and night (as well as in person). Many of them read and commented on my manuscript when surely they had better things to do. To Jamsheed Choksy, Paula Fredriksen, Peter Hawkins, Kathleen Flake, Martin Marty, Jeffrey Burton Russell, John Voll, N. T. Wright, and Carol Zaleski, I offer my boundless gratitude.

The process of writing a book is long and sometimes arduous, and through this period of purgation, Alan Segal and Stephen Prothero have become my friends. In addition to answering my persistent queries, both read the manuscript and delivered indispensable comments. Alan's expertise in the afterlife beliefs of the ancient world is unsurpassed, and his willingness to respond to my requests for translations and interpretations of Hebrew and Greek words—on one occasion instantly and from his hospital bed—has taken my breath away. Alan doesn't do his job, he lives it. Steve is a born teacher, and if his students receive a modicum of the enthusiasm and attention he lavished on me, they are fortunate indeed. Our conversations continue to be clarifying—and fun.

As the adage goes, "It takes a village." This book could not have been born without the indispensable help of David Gates and Ben Whitford. David is what they call "a writer's writer." His searing intelligence, his appetite for ideas, his desiccated wit—not to mention his fluency in literature and history—made him the perfect editor for this book. The day he agreed to help me was a lucky one. Ben Whitford came aboard early as a researcher, and then stayed on to do more research, fact checking, and footnoting. Thorough, exacting, organized, and full of excellent humor, Ben kept his cool even when the manuscript deadline coincided almost exactly with the birth of his first child. Any mistakes or errors of interpretation are entirely my own.

I am one of the few humans on the planet lucky enough to do what they love, every day, among people who are friends as well as colleagues. I continue to be grateful to Jon Meacham for having granted me this unique privilege. Jon believes fiercely in the importance of religion as a force in daily personal and political life, and he gave me the gift of being able to cover it for *Newsweek*. At *Newsweek* (past and present), for making my stories better and my weeks more fun, thanks to Dan Klaidman, Mark Miller, Ted Moncreiff, Bret Begun, Susannah Meadows, Carrie Levy, Vanessa Juarez, Adam Kushner, and Matthew Philips. Sally Quinn was a colleague first, but soon she became my friend, and my life is much richer for it.

Dorothy Kalins, always a force for good, edited the original cover story—which I couldn't have written without the help of Christopher Dickey, Anne Underwood, Joanna Chen, and Dan Ephron. Other editors and mentors contributed—wittingly and unwittingly—to this project. Mark Whitaker hired me at *Newsweek* and gave me every kind of support, including time to write. At the *Wall Street Journal*, Paul Steiger and Dan Hertzberg took a risk and offerred me the religion beat. Cynthia Crossen is probably the most gifted editor I've ever worked with: ten years on, I still hear her voice in my head.

Claire Wachtel at HarperCollins was passionate and enthusiastic from the outset—and then waited for this book with the patience of a (Jewish) saint; Julia Novitch helped the manuscript over last-minute bumps with grace. My agent, David McCormick, has been my friend for more than twenty years. He is a neighbor, a buddy, and a fierce advocate.

Friends who are also clear-eyed readers commented on drafts of the manuscript and in conversation helped me hash out problems along the way. These are Jerry Adler, Jonathan Darman, Jeff Giles, Rebecca Gradinger, and Laura Marmor. For the past fifteen-plus years, Kathy Deveny has served triple duty: editor, friend, confidante. "Wouldn't it be cool," she once said, "if you wrote a book about heaven?" Craig Townsend is a scholar, an Episcopalian priest, and a friend. He officiated at our interfaith wedding, he grills a mean steak, and he gave his full attention to the manuscript at the eleventh hour. His comments were invaluable.

Other friends gave ongoing support—not to mention wine, food, company, and hours on the telephone. To Ginia Bellafante, Elizabeth Causey, Rachel Cline, Virginia Genereux, Betsy Gleick, and Nancy Tompkins: Thanks, girls, as always. Joshua Patner belongs to this last group, too. John and Jake Dixon live in southern Vermont, in what can only be called Eden, and they, together with their daughter, Amanda, gave me a precious week of hospitality and writing time. Beverly Adjodha helps us keep it all together.

It is hard to imagine how to even go about thanking my family. One of the ideas of heaven I like best is that of continuity between the generations, born in biblical times and carried out not just in Judaism but in Roman Catholicism, Mormonism, and many other religions as well. I love the idea that a human being is connected to her ancestors and her children and children's children forever. And so, in the spirit of that eternal connectedness, I want especially to say a prayer to the memories of my four grandparents: Rita Waschman Goldmuntz and Marcel Goldmuntz on my mother's side; Florence Levy Miller and Irving George Miller on my father's. My own parents, Arlette and George Miller, gave me everything. To my mother, thank you for reading me stories from the Hebrew Bible instead of other children's books, which you deemed "too insipid." To my father, thank you for teaching me that to get anything done, you have to keep your butt in your chair. To John and Dave, thanks for being on the other end of the phone, and to Lauren, Tien-Yi, Beth, Wendy, Randy, Ellen, Kathee, and Millicent, thanks for doing all you did to keep our family life moving forward while I checked out for long, sometimes indeterminate periods of time.

My daughter, Josephine, reminds me every day that life is a miracle. And Charlie, who—literally and despite the cliché—knows me better than I know myself (and who has taken innumerable trips to the grocery store, the park, the museum, and the zoo, not to mention Detroit, California, Boston, and Vermont, so that I could have a few hours alone in the apartment), you are my heart. I am reminded, with gratitude, of what Craig said to us on our wedding day: the love between two imperfect people in marriage is the closest approximation we have on earth to the love of God.

NOTES

Unless otherwise indicated, all quotations from the Bible are from *The Holy Bible: Containing the Old and New Testaments with the Apocryphal/Deuterocanonical Books: New Revised Standard Version* (New York: Oxford University Press, 1989); and all quotations from the Qur'an are from Tarif Khalidi's verse translation: T. Khalidi, *The Qur'an* (London: Penguin Books, 2008).

INTRODUCTION

ix *A cover story*: L. Miller, "Why We Need Heaven," *Newsweek*, August 12, 2002.

xi *"An act against all"*: Address to the Greater Houston Ministerial Association, September 12, 1960. Quoted in R. V. Friedenberg, *Notable Speeches in Contemporary Presidential Campaigns* (Westport, CT: Praeger, 2002), p. 61.

xi *Nearly 80 percent*: The Pew Forum on Religion and Public Life, *U.S. Religious Landscape Survey*, 2008, p. 10.

xi *Freedom of religious expression*: Note, though, that these rights were not universal; Quakers, in particular, were often persecuted for their beliefs. See C. G. Pestana, *Quakers and Baptists in Colonial Massachusetts* (Cambridge: Cambridge University Press, 1991). For an excellent analysis of the founders and religion, see J. Meacham, *American Gospel: God, the Founding Fathers and the Making of a Nation* (New York: Random House, 2006).

xi *According to a 2007 poll*: Pew Forum on Religion and Public Life, *U.S. Religious Landscape Survey: Religious Affiliation*, pp. 5 and 174. See also their follow-up study, which found 65 percent believe many religions can lead to eternal life: The Pew Forum on Religion and Public Life, "Many Americans Say Other Faiths Can Lead to Eternal Life," 2008, http://pewforum.org/docs/?DocID=380.

xii *"The way our fathers took a medicine"*: Letter to Louis de Kergorlay, Yonkers, June 29, 1831, in A. de Tocqueville, *The Tocqueville Reader: A Life in Letters and Politics*, ed. O. Zunz and A S. Kahan (Malden, MA: Blackwell, 2002), pp. 44–45.

xii   *"We're no longer living"*: L. Miller, "Religion: The Age of Divine Disunity—Faith Now Springs from a Hodgepodge of Beliefs," *Wall Street Journal*, February 10, 1999.

xii   *A third of people say*: European Commission, *Special Eurobarometer 225: Social Values, Science and Technology*, Brussels, Belgium, 2005, p. 9.

xii   *We retain our religiosity*: Pew Forum on Religion and Public Life, *Religious Landscape Survey*, p. 162.

xii   *In sixty years*: A. L. Winseman, "Eternal Destinations: Americans Believe in Heaven, Hell," Gallup, Inc., May 25, 2004, http://www.gallup.com/poll/11770/Eternal-Destinations-Americans-Believe-Heaven-Hell.aspx.

xii   *Losing adherents*: Pew Forum on Religion and Public Life, *Religious Landscape Survey*, p. 26.

xiii *So when 81 percent*: Gallup, Inc., "Gallup/Nathan Cummings Foundation and Fetzer Institute Poll," Gallup, Inc., May 10–13, 2007, and May 1997, http://www.gallup.com/poll/1690/Religion.aspx.

xiv   *The most successful denominations*: "World Christian Database," Gordon-Conwell Theological Seminary, 2005, http://www.worldchristiandatabase.org.

xiv   *Living the right kind of life*: A turn of phrase I take to include both doing good works and submitting to God's mysterious will.

xiv   *Heaven changed during the Civil War*: P. S. Paludan, "Religion and the American Civil War," in *Religion and the American Civil War*, ed. R. M. Miller, H. S. Stout, and C. R. Wilson (New York: Oxford University Press, 1998), p. 30.

xv   *"Walk off"*: A. G. Lotz, *Heaven: My Father's House* (Nashville, TN: W Publishing Group, 2001), p. 47.

xvi   *Orthodox apologists*: The word orthodox means "right belief," and I've used it throughout to refer to conservative, traditionally observant religious groups; note, however, that many progressives also describe their religious beliefs as orthodox.

xvii *"Not just words"*: L. Miller and R. Wolffe, "I Am a Big Believer in Not Just Words, But Deeds and Works," *Newsweek*, July 21, 2008.

xviii *The town of Ripley*: K. P. Griffler, *Front Line of Freedom: African Americans and the Forging of the Underground Railroad in the Ohio Valley* (Lexington: University Press of Kentucky, 2004), pp. 61–62.

xviii *"A yellow Cadillac convertible"*: W. C. Martin, *A Prophet with Honor: The Billy Graham Story* (New York: W. Morrow, 1991), p. 126.

xviii *"A whole series of symbols"*: See Reinhart's introduction to L. Bakhtiar, *Encyclopedia of Islamic Law: A Compendium of the Views of the Major Schools* (Chicago: ABC International Group, 1996).

xix   *81 percent of Americans*: Gallup, Inc., Gallup-Nathan Cummings Foundation and Fetzer Institute Poll.

xix   *"Heaven is AWOL"*: D. Van Biema, "Does Heaven Exist?" *Time*, March 24, 1997.

xx   *"Tinky Winky"*: J. Falwell, "Tinky Winky Comes Out of the Closet," NLJ Online, February, 1999, http://web.archive.org/web/19990423025753/; http://www.liberty .edu/chancellor/nlj/feb99/politics2.htm.

xxi *"Entering the happiest life"*: B. Woodward, "In Hijacker's Bags, a Call to Planning, Prayer and Death," *Washington Post*, September 28, 2001.

xxi *By journalists like Paul*: Barrett ended up writing a book about the Muslim experience in the United States. P. Barrett, *American Islam: The Struggle for the Soul of a Religion* (New York: Farrar, Straus and Giroux, 2007).

xxii *"Boughton says he has more ideas"*: M. Robinson, *Gilead* (New York: Farrar, Straus and Giroux, 2004), p. 147. Thanks to Peter Hawkins, at Yale University, for reminding me of this beautiful passage.

xxiii *"People have been told so often"*: J. Meacham and L. Miller, "Everything Old Is New Again," *Newsweek*, May 5, 2008.

xxiv *"Unhappy America"*: *Economist*, July 24, 2008.

xxiv U.S. Census Bureau, "Interim Projections of the Population by Selected Age Groups for the United States and States: April 1, 2000, to July 1, 2030," April 21, 2005, http://www.census.gov/population/projections/SummaryTabB1.pdf.

xxiv *A 2003 Pew poll*: Pew Research Center for People and the Press, *2004 Political Landscape*, November 5, 2003, http://people-press.org/report/196.

xxiv *"Flying spaghetti monster"*: S. Harris and A. Sullivan, "Is Religion 'Built Upon Lies'?" BeliefNet, January 23, 2007, http://www.beliefnet.com/Faiths/Secular-Philosophies/Is-Religion-Built-Upon-Lies.aspx. The Flying Spaghetti Monster was conceived in 2005 by Bobby Henderson, then an unemployed slot-machine engineer, as a parody of the arguments made by proponents of Intelligent Design; see B. Henderson, "Open Letter to Kansas School Board," Church of the Flying Spaghetti Monster, January 2005, http://www.venganza.org/about/open-letter/.

xxiv *"Suspension of disbelief"*: S. T. Coleridge, *Biographia Literaria* (London: G. Bell and Sons, 1898), p. 145.

xxvi *This feeling of reverence*: Not everyone thinks of God in this way, of course. The theologian N. T. Wright notes that to some believers, my description of God as a "feeling of reverence" seems to miss the point. "If I said 'this feeling of warmth and snug comfort is what I have always called my wife,' I think my wife might hit me," he adds (e-mail, April 21, 2009).

CHAPTER ONE: WHAT IS HEAVEN?

2 *A spiritual memoir*: J. Martin, *My Life with the Saints* (Chicago: Loyola Press, 2006).

3 *"If you have to explain it"*: D. Goodyear, "Quiet Depravity," *New Yorker*, October 24, 2005.

3 *Saints populate heaven*: For a broader discussion of the role of the saints in Catholic theology, see P. R. L. Brown, *The Cult of the Saints: Its Rise and Function in Latin Christianity* (Chicago: University of Chicago Press, 1981).

3 *"I believe in hell"*: E. Schillebeeckx, *For the Sake of the Gospel* (New York: Crossroad, 1990), p. 111.

4 *God sent them to earth*: Abraham's visitation is described in Genesis (22:11); Moses's in Exodus (3:2); and Mary's in Luke (1:26–38).

4 *"For beauty is nothing"*: "Elegy 1," trans. Stephen Mitchell, quoted in W. H Gass, *Reading Rilke: Reflections on the Problems of Translation* (New York: Alfred A. Knopf, 1999), pp. 65–66.

4 *Pearl gates*: Revelation 21:18–21. Note that some scholars advocate an earlier date, between the mid-50s and early 70s CE, for Revelation's authorship. See G. Desrosiers, *An Introduction to Revelation* (London: Continuum International Publishing Group, 2000), p. 50.

5 *"I will give you the keys"*: Matthew 16:19.

5 *"Eat pâté de foie gras"*: C. Arnold-Baker, *The Companion to British History* (London: Routledge, 2001), p. 1148.

5 *"You'll behold the Union Depot"*: "Life's Railway to Heaven," copyrighted 1890 by M. E. Abbey and Charles D. Tillman, quoted in N. Cohen and D. Cohen, *Long Steel Rail: The Railroad in American Folksong* (Urbana: University of Illinois Press, 2000), p. 612.

5 *A 2002 Newsweek poll*: L. Miller, "Why We Need Heaven," *Newsweek*, August 12, 2002.

6 *"First-century conceptuality of heaven"*: E-mail from N. T. Wright, April 21, 2009.

6 *Different word for each heavenly concept*: I am especially indebted to Alan Segal for his translations and interpretations of the Hebrew Bible.

7 *"The gray rain-curtain of this world"*: The words are adapted from Tolkien's original text, in which the narrator describes Frodo's experience of leaving Middle Earth for the "undying land" beyond the sea. See J. R. R. Tolkien, *The Return of the King: Being the Third Part of The Lord of the Rings* (New York: Random House, 1983), p. 339.

8 *"Let the little children"*: Matthew 19:14.

8 *"They lived in great joy"*: C. S. Lewis, *The Lion, the Witch and the Wardrobe; a Story for Children* (New York: Macmillan, 1950), p. 169.

8 *"All their life in this world"*: C. S. Lewis, *The Last Battle* (New York: HarperCollins, 1994), p. 228.

8 *31 percent of people*: The Pew Forum on Religion and Public Life, *U.S. Religious Landscape Survey: Religious Affiliation*, Washington, DC, 2008.

9 *The Dead Sea Scrolls*: The majority scholarly view is that the Essenes, or a group very much like the Essenes, wrote the Dead Sea Scrolls. See, for example, J. H.

Charlesworth, *Jesus and the Dead Sea Scrolls* (New Haven, CT: Anchor Bible Series, 1995); and J. Magness, *The Archaeology of Qumran and the Dead Sea Scrolls* (Grand Rapids, MI: William B. Eerdmans, 2002). A minority view, that the Essenes did not write the scrolls—or perhaps never existed—is represented by R. Elior, *Memory and Oblivion: The Secret of the Dead Sea Scrolls* (Jerusalem: Van Leer Institute and ha-Kibutz ha-Meuchad, 2009).

10 *Showed Jesus ascending*: R. Deshman, "Another Look at the Disappearing Christ: Corporeal and Spiritual Vision in Early Medieval Images," *The Art Bulletin* 79, no. 3 (1997): 518–46.

12 *They had a deity called Ba'al*: For a complete description of Canaanite afterlife beliefs, see A. F. Segal, *Life After Death: A History of the Afterlife in the Religions of the West* (New York: Doubleday, 2004), pp. 104–109; see also A. R. W. Green, *The Storm-God in the Ancient Near East* (Winona Lake, IN: Eisenbrauns, 2003), p. 222.

12 *Osiris, who lived among the stars*: Alan Segal writes, "Osiris lives in the heavens and underworld. Both at once and I don't know how. The constellation Orion is called the house of Osiris in Egyptian. In Winter Orion is high in the sky but in the summer part of it sets." (By e-mail, August 6, 2009.) Osiris was believed to have the power to grant both earthly fertility and eternal life. See D. H. Kelley and E. F. Milone, *Exploring Ancient Skies: An Encyclopedic Survey of Archaeoastronomy* (New York: Springer, 2005), p. 261; see also B. Mosjov, *Osiris: Death and Afterlife of a God* (Malden, MA: Blackwell, 2005).

12 *"Possessor of heaven"*: Genesis 14:18–22, in *The Holy Bible (King James Version)* (New York: Cambridge University Press, 2000).

13 *"At the blast of your nostrils"*: Exodus 15:1–20.

13 *"Corn from heaven"*: Exodus 16; Numbers 11:7–9; Psalms 78:24.

13 *The predominant view of cosmology*: E. Grant, *Physical Science in the Middle Ages* (Cambridge: Cambridge University Press, 1977), pp. 71–72.

13 *Rabbi Ishmael*: Segal, *Life After Death*, p. 513.

13 *Ascends through seven spheres*: Ibid., pp. 654–55. See also F. S. Colby, *Narrating Muhammad's Night Journey: Tracing the Development of the Ibn 'Abbâs Ascension Discourse* (Albany: State University of New York Press, 2008), pp. 1–2.

14 *In Christian cosmology*: R. Hughes, *Heaven and Hell in Western Art* (London: Weidenfeld & Nicolson, 1968), pp. 112–14. This model is cast by some scholars as a Dantean innovation; see A. Cornish, "Angels: Number and Hierarchy," in R. H. Lansing, ed., *The Dante Encylopedia* (New York: Garland, 2000), pp. 42–43. For a more complete discussion of the adaptation of Aristotelian cosmology by Christian theologians, and particularly the installation of the immobile Empyrean sphere beyond Aristotle's primum mobile, see E. Grant,

Planets, Stars, and Orbs: The Medieval Cosmos, 1200–1687 (Cambridge: Cambridge University Press, 1994), pp. 371–421.

14 *"Light that flowed"*: Canto 30:62–63 in Dante Alighieri, *Paradiso*, verse translation by Robert and Jean Hollander, introduction and notes by Robert Hollander (New York: Doubleday, 2007), p. 735.

14 *Around the year 200 BCE: The Encyclopedia of Apocalypticism*, ed. B. McGinn, J. J. Collins, and S. J. Stein (New York: Continuum, 1998). See also *Frontline*, "Apocalypse!" first broadcast November 22, 1998, by PBS. Written, produced, and directed by W. Cran and W. Loeterman; and associated primary-source materials: L. M. White, "Apocalyptic Literature in Judaism and Early Christianity," PBS, 1998, http://www.pbs.org/wgbh/pages/frontline/shows/apocalypse/primary/white.html.

15 *Post–9/11 jihadis*: R. Paz, "Hotwiring the Apocalypse: Apocalyptic Elements of Global Jihadi Doctrines," in M. Sharpe (ed.), *Proceedings of the NATO Advanced Research Workshop on Suicide Bombers—the Psychological, Religious and Other Imperatives* (Cambridge: IOS Press, 2008,) pp. 108–109.

15 *Fundamentalist Christians*: L. Miller, "Is Obama the Antichrist?" *Newsweek*, November 15, 2008.

15 *"God himself will be with them"*: Revelation 21:3–4.

15 *Minority view*: Some scholars strongly disagree with Slonim's suggestion that few Lubavitchers hold the Rebbe to have been the Messiah. For a full account of the opposing view, see D. Berger, *The Rebbe, the Messiah, and the Scandal of Orthodox Indifference* (London: Littman Library of Jewish Civilization, 2001).

16 *"Knowledge of God"*: See also Isaiah 11:9.

17 *19 percent of Americans*: Miller, "Why We Need Heaven."

17 *Christianity was brand-new*: According to an early tradition, Christians—forewarned by Christ—fled Jerusalem around 66 CE, probably to Pella in the Transjordan Valley, and thus weren't present for the destruction of the city. Still, modern scholars say there's little historical evidence to support the story, and that Christians likely remained in and around Jerusalem until its destruction. See L. Michael White, *From Jesus to Christianity* (San Francisco: HarperSanFrancisco, 2004), pp. 229–31.

17 *13 percent*: Miller, "Why We Need Heaven."

17 *A world of farmers*: Segal, *Life After Death*, p. 105.

17 *Garden walls*: Hughes, *Heaven and Hell*, p. 47.

18 *"Every tree"*: Genesis 2:9–19.

18 *Monks drew maps*: A. Scafi, *Mapping Paradise: A History of Heaven on Earth* (Chicago: University of Chicago Press, 2006).

18 *"I believe that in the earthly paradise"*: As quoted in J. Delumeau, *History of Para-*

dise: The Garden of Eden in Myth and Tradition, trans. Matthew O'Connell (New York: Continuum, 1995), p. 54.

18 *"'Heaven'—is what I cannot Reach!"*: E. Dickinson, *The Poems of Emily Dickinson: Reading Edition*, ed. R. W. Franklin (Cambridge, MA: Belknap Press of Harvard University Press, 1999), p. 137.

19 *An appropriation by Christian painters*: R. M. Jensen, *Understanding Early Christian Art* (London: Routledge, 2000), p. 112.

19 *Eden became a popular subject*: C. McDannell and B. Lang, *Heaven: A History* (New York: Vintage Books, 1990), pp. 111–12.

19 *The Blessed Virgin*: The central figure in La Primavera is more generally assumed to be Venus, surrounded by Cupid, Mercury, and the Graces. For an overview of the consensus position—and of the case for a Marian reading—see K. A. Lindskoog, "Botticelli's 'Primavera' and Dante's 'Purgatory,' " in *Dante's Divine Comedy: Purgatory* (Macon, GA: Mercer University Press, 1997), ix–xiv.

19 *Gozzoli's fifteenth-century fresco*: See the analysis in A. E. McGrath, *A Brief History of Heaven* (Malden, MA: Blackwell, 2003), p. 58.

19 *"A most pleasant garden"*: L. de Medici and L. Cavalli, *Opere: A Cura di Luigi Cavalli* (Naples: F. Rossi, 1970), p. 368, translated in E. B. MacDougall, *Medieval Gardens* (Washington, DC: Dumbarton Oaks Research Library and Collection, 1986), p. 238. Quoted in McGrath, *A Brief History of Heaven*, p. 58.

20 *A safe and fertile place*: Segal, *Life After Death*.

20 *"There flowers of gold shine"*: From "Olympian 2," in A. Pindar, *The Complete Odes*, trans. A. Verity, with an introduction and notes by S. Instone (Oxford: Oxford University Press, 2007), p. 9.

21 *The idea of heaven as a paradise garden*: For more on the imagery of the Islamic paradise, see S. Blair and J. Bloom, eds., *Images of Paradise in Islamic Art* (Hanover, NH: Hood Museum of Art, Dartmouth College, 1991). Though images of the Prophet are forbidden in most Muslim traditions, the wealthy and powerful have, throughout history, built opulent gardens as reflections of Paradise. See D. F. Ruggles, *Islamic Gardens and Landscapes* (Philadelphia: University of Pennsylvania Press, 2008).

21 *They will inhabit . . . pomegranates*: See suras 2 ("The Cow"), verse 24; 76 ("Man"), verses 3–10; 47 ("Muhammad"); and 55 ("The All-Merciful").

21 *In a pomegranate garden*: C. G. Jung, *Memories, Dreams, Reflections* (New York: Vintage Books, 1989), p. 294.

21 *Blood and . . . fertility*: F. J. Simoons, *Plants of Life, Plants of Death* (Madison: University of Wisconsin Press, 1998), p. 279.

21 *In the Muslim paradise, according to the Qur'an*: The imagery described in this

paragraph can be found in suras 56 ("The Event"); 76 ("Man"); 55 ("The All-Merciful"); and 15 ("Al-Hijr").

22 *"I have sometimes suggested"*: T. Carnes and A. Karpathakis, eds., *New York Glory: Religions in the City* (New York: New York University Press, 2001), p. xiii.

22 *Jerusalem was the center*: See M. Goodman, *The Ruling Class of Judaea: The Origins of the Jewish Revolt Against Rome, A.D. 66–70* (Cambridge: Cambridge University Press, 1987).

22 *To slay his beloved son*: Genesis 22:1–14.

23 *"The holy city Jerusalem"*: Revelation 21:10.

23 *Jeweled walls*: Revelation 4:6.

23 *A mosaic from 440 CE*: Thanks to Ena Heller, of the Museum of Biblical Art, for showing me this beautiful mosaic.

23 *"A challenge and an invitation"*: A. Pilla, "Building the City of God," EcoCity Cleveland, November, 1993, from http://www.ecocitycleveland.org/smartgrowth/cornfields/city_of_god.html.

23 *"I dream of someday"*: M. Brown, "About the Author," LatterDayLogic.com, 2007, from http://www.latterdaylogic.com/about/.

23 *Gingerich is an astrophysicist*: See also his book, O. Gingerich, *God's Universe* (Cambridge, MA.: Belknap Press of Harvard University Press, 2006).

CHAPTER TWO: THE MIRACLE

28 *Jews who invented our idea of heaven*: Here there is no scholarly consensus. Most scholars agree that Daniel 12:2–3 gave resurrection and eternal life to some people, some of the time. They disagree vehemently, however, on what the verse actually means. In the resurrection, will humans be humans or something more like angels? And do the seeds of modern-day heaven, cozy and populated as it is with our friends and relatives, lie in the ancient idea of resurrection? For a range of views, see N. Gillman, *The Death of Death: Resurrection and Immortality in Jewish Thought* (Woodstock, VT: Jewish Lights, 1997); J. J. Collins, *Daniel: With an Introduction to Apocalyptic Literature* (Grand Rapids, MI: William B. Eerdmans, 1984); and A. F. Segal, *Life After Death: A History of the Afterlife in the Religions of the West* (New York: Doubleday, 2004).

28 *The Jewish patriarch Abraham*: For the stories of Sarah's burial and of Abraham's death, see Genesis 23:2–20 and 25:8–10.

29 *Not the dwelling of humans*: I am indebted to Alan Segal for his readings and interpretations of the Hebrew Bible here.

29 *Full of allusions*: Noah's flood, Genesis 6:9–8:22; the story of Lot, Genesis 19:24; Hagar and Ishmael, Genesis 21:17; Abraham and Isaac, Genesis 22:11.

29 *Jacob lays his head down*: Genesis 28:11–17.

30 *Babel*: Genesis 11:4–9.

30 *When Jacob dies*: See Genesis 49:29–30 and 50:2–13.

30 *Buried in caves*: Thanks to Rachel Hallote, director of the Jewish Studies Program at SUNY-Purchase, and Jodi Magness, Kenan Distinguished Professor for Teaching Excellence in Early Judaism at UNC-Chapel Hill, for their help explaining ancient Israelite burial customs. See also E. Bloch-Smith, *Judahite Burial Practices and Beliefs About the Dead* (Sheffield, UK: JSOT Press, 1992).

30 *Those caves are full of bones*: In 2007, the Canadian filmmaker Simcha Jacobovici and his coauthor Charles Pellegrino published a book in which they claimed to have found a tomb near Jerusalem containing the bones of members of Jesus's family and perhaps even of Jesus himself. The book, and a television program linked to its publication, caused a massive and not entirely favorable stir. S. Jacobovici and C. Pellegrino, *The Jesus Family Tomb: The Discovery, the Investigation, and the Evidence That Could Change History* (New York: HarperOne, 2007).

31 *"So take the past"*: From Darwish's 1988 poem "Those Who Pass Between Fleeting Words," quoted in Z. Lockman and J. Beinin, *Intifada: The Palestinian Uprising Against Israeli Occupation* (Boston: South End Press, 1989), pp. 26–27.

32 *"This is my covenant"*: Genesis 17:4–8.

32 *"Take away our disgrace"*: Isaiah 4:1.

32 *Went to Sheol*: Scholars are still debating the precise role played by Sheol in early Judaism. Some, like Jon Levenson at Harvard University, believe that Sheol was a destination for people who died in dishonor or were in some other way unfulfilled. Alan Segal argues that everyone went to Sheol, and not just those who displeased God.

32 *"As the cloud fades"*: Job 7:9–10.

34 *The Hebrews lived in multigenerational*: Descriptions in this paragraph come largely from R. S. Hallote, *Death, Burial, and Afterlife in the Biblical World: How the Israelites and Their Neighbors Treated the Dead* (Chicago: Ivan R. Dee, 2001).

35 *And indeed many other scholars*: See, for example, Bloch-Smith, *Judahite Burial Practices*. Also note R. Hachlili, *Ancient Jewish Art and Archaeology in the Land of Israel* (Leiden: E. J. Brill, 1988); and R. E. Friedman, *Who Wrote the Bible?* (New York: Summit Books, 1987).

35 *"No one shall be found"*: Deuteronomy 18:10–11.

35 *"Stoned with stones"*: Leviticus 20:27.

35 *The story of King Saul*: Samuel 28:3–19.

37 *The Bible says*: Out of slavery, Exodus 12:41; forty years in the desert, Exodus 16:35; Ten Commandments, Deuteronomy 5:1–22; Moses died, Deuteronomy

34:1–5; back to Canaan, Joshua 1–13; King David, 1 Chronicles 11:4–7; Abraham and Isaac, 2 Chronicles 3:1.

38 *Drove many of the people*: B. Porten, "Exile, Babylonian," in *Encyclopaedia Judaica*, ed. M. Berenbaum and F. Skolnik (Detroit: Macmillan Reference USA, 2007), vol. 6, pp. 608–11.

38 *The great Persian Empire*: G. R. Garthwaite, *The Persians* (Malden, MA: Blackwell, 2004).

38 *Nile River island*: A. Schalit and L. Matassa, "Elephantine," in *Encyclopaedia Judaica*, ed. Berenbaum and Skolnik, vol. 6, pp. 311–14. Thanks also to Jamsheed Choksy, Professor of Central Eurasian Studies at Indiana University-Bloomington, for his help on the Exile.

39 *Began to create a religion*: R. Drews, "Judaism, Christianity, and Islam to the Beginnings of Modern Civilization," Vanderbilt University, April 20, 2009, http://site mason.vanderbilt.edu/classics/drews/COURSEBOOK, chap. 4.

39 *Coincided with . . . a new religion*: Segal, *Life After Death*. See also R. Stark, *Discovering God: The Origins of the Great Religions and the Evolution of Belief* (New York: HarperOne, 2007), pp. 188–191; and P. Clark, *Zoroastrianism: An Introduction to an Ancient Faith* (Brighton: Sussex Academic Press, 1999). Some scholars put less weight on Zoroastrian influences, noting the paucity of contemporary documentary evidence pertaining to Zoroastrian beliefs and practices at the time of the Exile. See N. T. Wright, *The Resurrection of the Son of God* (Minneapolis: Fortress Press, 2003).

39 *Zoroastrian scripture*: Segal, *Life After Death*, pp. 189.

40 *"I form light"*: Isaiah 45:7.

40 *There are not two Gods*: Segal, *Life After Death*, pp. 198–200. See also L. A. Hoffman, *The Journey Home: Discovering the Deep Spiritual Wisdom of the Jewish Tradition* (Boston: Beacon Press, 2002), p. 143. Note, however, that this view is not universally accepted; Choksy, in particular, argues that Jews and Zoroastrians would have seen one another as monotheistic.

41 *Jerusalem wasn't much of a place*: Thanks to Hanan Eshel, former head of the Department of Land of Israel Studies and Archaeology at Bar-Ilan University, Israel, for his help via phone and e-mail reconstructing the mood and geography of Jerusalem in the centuries following the Exile and leading up to the Maccabean Revolt.

41 *Beginning to change*: V. Tcherikover, *Hellenistic Civilization and the Jews* (Philadelphia: Jewish Publication Society of America, 1959).

41 *Jacob was mummified*: Genesis 50:2–3.

42 *Torah metes out punishment*: Deuteronomy 18:10–12; see also Leviticus 20:27.

42 *"Enoch walked with God"*: Genesis 5:17–24.

42 *Genesis . . . written and finalized*: T. L. Brodie, *Genesis as Dialogue: A Liter-*

ary, Historical, and Theological Commentary (Oxford: Oxford University Press, 2001), pp. 80–85. See also R. E. Friedman, *Who Wrote the Bible?*

42 *"Enoch" himself*: Note that the first book of Enoch was considered scriptural by the authors of the New Testament—it's quoted as true scripture in Jude 14—and remains canonical for Ethiopian Christians. For more detail, see D. C. Olson, "1 Enoch," in *Eerdmans Commentary on the Bible*, ed. J. D. G. Dunn and J. W. Rogerson (Grand Rapids, MI: William B. Eerdmans, 2003), pp. 904–41.

43 *Elijah, the prophet . . . "taken up"*: 1 Kings 17–21 and 2 Kings 2:11.

43 *Some Jews, some of the time*: I am indebted to Mark Smith, Professor of Hebrew and Judaic Studies at New York University, for his insights about Enoch and Elijah. See also M. Smith, *The Memoirs of God: History, Memory and the Experience of the Divine in Ancient Israel* (Minneapolis: Fortress Press, 2004); and A. Amanat and M. T. Bernhardsson, *Imagining the End: Visions of Apocalypse from the Ancient Middle East to Modern America* (London: I. B. Tauris, 2002), pp. 72–78.

43 *Influence . . . of the Greeks*: Segal, *Life After Death*.

43 *Jerusalem had been gobbled up*: A. Schalit, E. E. Halevy, J. Dan, and A. Saenz-Badillos, "Alexander the Great," in *Encyclopaedia Judaica*, ed. M. Berenbaum and F. Skolnik (Detroit: Macmillan Reference USA, 2007), vol. 1, pp. 625–27.

44 *In the third century BCE*: Tcherikover, *Hellenistic Civilization*. See also J. H. Hayes and S. Mandell, *The Jewish People in Classical Antiquity* (Louisville, KY: Westminster John Knox Press, 1998), pp. 5–6.

45 *Alexander's own favorite sculptor*: "Lysippos," in *The Grove Encyclopedia of Classical Art and Architecture* (e-reference ed.), ed. G. Campbell (Oxford University Press, 2007), www.oxfordreference.com/views/ENTRY.html?subview=Main&entry=t231.e058.

45 *The Greeks believed in the soul*: J. N. Bremmer, *The Early Greek Concept of the Soul* (Princeton, NJ: Princeton University Press, 1983), pp. 16–17.

45 *"Departs to the invisible world"*: Plato, *Phaedo* (Stillwell, KS: Digireads.com, 2006), p. 58.

45 *Plato believed*: Note that it's unclear whether Plato intended his discussion of the Demiurge to be taken literally; generations of Plato's followers and critics have argued the case one way and the other. See R. D. Mohr, *God and Forms in Plato: The Platonic Cosmology* (Las Vegas: Parmenides Publishing, 2005).

46 *Believed in reincarnation*: Segal, *Life After Death*, pp. 232–34. See also M. R. Taylor, "Dealing with Death: Western Philosophical Perspectives," in *Handbook of Death and Dying*, ed. C. D. Bryant, 1:24–33 (Thousand Oaks, CA: Sage, 2003).

46 *Tensions began to emerge*: Tcherikover, *Hellenistic Civilization*. See also J. J. Collins, *Daniel: With an Introduction to Apocalyptic Literature* (Grand Rapids, MI: William B. Eerdmans, 1984).

46 *The same question*: E. J. Bickerman and Jewish Theological Seminary of America, *The Jews in the Greek Age* (Cambridge, MA: Harvard University Press, 1988).

47 *Jerusalem's high priest*: U. Rappaport, "Jason," in *Encyclopaedia Judaica*, ed. Berenbaum and Skolnik, vol. 11, p. 90.

48 *Goings-on at the gymnasium*: 2 Maccabees 4:12 and 1 Maccabees 1:15.

48 *The story of the Jewish revolt*: Tcherikover, *Hellenistic Civilization*.

48 *Menelaus paid the king*: 2 Maccabees 4:32.

49 *According to the books of the Maccabees*: See especially 1 Maccabees 1:41–60 and 2 Maccabees 5:27, 8:16, and 8:7.

49 *A man whom I will call Daniel*: Like many of the Bible's books, the book of Daniel is believed to have many authors, and to have been assembled over a period of many years; nonetheless, the apocalyptic verses referenced here can be confidently dated to the time of the Maccabean revolt. For a more complete discussion, see Collins, *Daniel*.

49 *"Many of those who sleep"*: Daniel 12:2–3.

CHAPTER THREE: THE KINGDOM IS NEAR

53 *"Has already been taken up"*: M. Shriver, *What's Heaven?* (New York: Golden Books, 1999), p. 19.

54 *61 percent of Americans*: *Time*/CNN poll conducted March 11–12, 1997, by Yankelovich Partners, Inc. See D. Van Biema, "Does Heaven Exist?" *Time*, March 24, 1997.

54 *"The new heaven"*: See Revelation 21:1.

54 *In the Gospel of Matthew*: The farmer, the mustard seed, the fishing net, and the pearls can be found in Matthew 13:24–46; the king's accounts in Matthew 18:23–35; the king's banquet in Matthew 22:1–14; the landowner in Matthew 20:1–16; the ten virgins in Matthew 25:1–13; the camel and the needle in Matthew 19:23–24; the warning against earthly wealth in Matthew 6:19–21; the command to become as children in Matthew 18:3; and the warning against ostentatious piety in Matthew 6:1.

55 *"The time is fulfilled"*: Mark 1:14–15.

55 *He was posing a question*: N. T. Wright, e-mail dated April 21.

55 *"A final cosmic catastrophe"*: A. Schweitzer and W. Lowrie, *The Mystery of the Kingdom of God: The Secret of Jesus' Messiahship and Passion* (London: A. & C. Black, 1925).

56 *Apocalyptic literature or scripture*: A great compendium of apocalyptic movements, from ancient times to the present day, is this one: *The Encyclopedia of Apocalypticism*, ed. B. McGinn, J. J. Collins, and S. J. Stein (New York: Continuum, 1998).

56 *Store cans of food*: L. J. Arrington and D. Bitton, *The Mormon Experience: A His-*

tory of the Latter-Day Saints, 2nd ed. (Urbana: University of Illinois Press, 1992), p. 277.

56 *A renewed planet*: M. J. Penton, *Apocalypse Delayed: The Story of Jehovah's Witnesses*, 2nd ed. (Toronto: University of Toronto Press, 1997), pp. 179–81.

57 *Simply boilerplate*: L. Miller and A. Murr, "Jesus and Witches," *Newsweek*, October 28, 2008, http://www.newsweek.com/id/166215.

57 *Apocalyptic expectation*: See, for instance, J.-P. Filiu, *L'apocalypse dans l'Islam* (Paris: Fayard, 2008).

57 *"The Hour is coming"*: Sura 20 *("Ta' Ha' ")*: 15.

57 *The destruction of Israel*: The necessity of the destruction of Israel, though preached by many Twelvers, isn't inherent to the idea of Mahdism; the scriptural justification for the position could as easily be taken to refer to the overthrow of Arab regimes or the implementation of a new world order. For more detail, see T. R. Furnish, *Holiest Wars: Islamic Mahdis, Their Jihads, and Osama bin Laden* (Westport, CT: Praeger, 2005). Says Jamsheed Choksy, professor of Central Eurasian Studies at Indiana University, Bloomington: "If Ahmadinejad et al. really were expecting the Mahdi to reappear during their lifetimes, or if they intended to trigger apocalyptic events, then their beliefs would require them to engage in preparing themselves through words and actions—including helping fellow Muslims, being honest, and purifying themselves. I don't see them doing any of that." (By e-mail, August 20, 2009.)

57 *"The real savior"*: Remarks to the United Nations General Assembly in New York, September 19, 2006, available at "Transcript of Ahmadinejad's U.N. Speech," NPR.org, September 19, 2006, http://www.npr.org/templates/story/story.php?storyId=6107339.

57 *Israel is similarly important*: P. S. Boyer, *When Time Shall Be No More: Prophecy Belief in Modern American Culture* (Cambridge, MA: Belknap Press of Harvard University Press, 1992), pp. 185–93.

57 *Hagee . . . McCain*: N. Guttman, "McCain Battles for Credibility with Jews," *The Jewish Chronicle* (Washington, DC), March 14, 2008. In 2006, Hagee told Reuters that "our support for Israel has absolutely nothing to do with an end times prophetic scenario."

58 *Claim their view as his own*: For an excellent analysis of the ways different American interest groups claim Jesus as their own, see S. Prothero, *American Jesus: How the Son of God Became a National Icon* (New York: Farrar, Straus and Giroux, 2003).

58 *"Ruling of God in our hearts"*: H. Pope, "The Kingdom of God," in *The Catholic Encyclopedia*, ed. C. G. Herbermann, E. A. Pace, C. B. Pallen, T. J. Shahan, and J. J. Wynne (New York: The Encyclopedia Press, 1913), vol. 8, pp. 646–47.

58 *No preoccupation with the hereafter*: J. D. Crossan, *Jesus: A Revolutionary Bi-*

ography (San Francisco: HarperSanFrancisco, 1994). See also J. D. Crossan, *The Historical Jesus: The Life of a Mediterranean Jewish Peasant* (San Francisco: HarperSanFrancisco, 1991).

58 *"Your Kingdom come"*: Matthew 6:9–13.

59 *watered-down apocalypticism*: See E. P. Sanders, *The Historical Figure of Jesus* (London: Allen Lane, 1993); E. P. Sanders, *Jesus and Judaism* (Philadelphia: Fortress Press, 1985); and P. Fredriksen, *From Jesus to Christ: The Origins of the New Testament Images of Jesus* (New Haven, CT: Yale University Press, 1988).

59 *Influenced by the Essenes*: J. E. Taylor, *The Immerser: John the Baptist Within Second Temple Judaism* (Grand Rapids, MI: William B. Eerdmans, 1997), pp. 15–48.

59 *Awareness . . . of the imminent end*: See, for example, Fredriksen, *From Jesus to Christ*.

59 *"You are the Messiah"*: Mark 8:27–30. See also Matthew 16:17–28.

60 *Centuries of speculation*: R. Bultmann, "The Message of Jesus and the Problem of Mythology," in *The Historical Jesus in Recent Research*, ed. J. D. G. Dunn and S. McKnight (Winona Lake, IN: Eisenbrauns, 2005), pp. 531–42. For an opposing view, see Wright, *Victory of God*.

60 *"For you yourselves know"*: 1 Thessalonians 5:2–4.

60 *"Into the invisible place"*: Irenaeus, "Against Heresies," trans. A. Roberts and W. Rambaut, in *Ante-Nicene Fathers*, vol. 1, ed. A. Roberts, J. Donaldson, and A. C. Coxe (Buffalo, NY: Christian Literature Publishing Co., 1885), p. 560.

60 *"This fire"*: "Ita plane quamuis salui per ignem, gravior tamen erit ile ignis, quam quidquid potest homo pati in hac vita" (Enarrationes in Psalmos 38.3 CCL, 38.384). Translation from J. Le Goff, *The Birth of Purgatory* (Chicago: University of Chicago Press, 1984), p. 69.

60 *"A never-ending present"*: Augustine, *Confessions* (Oxford: Clarendon Press, 1992), book 11, chapter 13, quoted in J. S. Feinberg, *No One Like Him: The Doctrine of God* (Wheaton, IL: Crossway Books, 2006), p. 381.

61 *The traditional Christian view*: Sparkling walls and gates of pearl are in Revelation 21:19–21; the tree of life is in Revelation 22:2; the light of God is in Revelation 22:5.

61 *"And let everyone who hears"*: Revelation 22:17.

62 *A third of America's white evangelicals*: The Pew Research Center for the People and the Press, "Many Americans Uneasy with Mix of Religion and Politics," www.pewforum.org/docs/?DocID-153.

62 *Reign . . . for a thousand years*: According to premillennialists, at least, who— like Lotz—believe the Second Coming will usher in a thousand years of peace. Some other Christians, called postmillennialists, believe that we are already

experiencing the rule of Christ; others still view the language describing Christ's thousand-year reign as symbolic rather than literal.

63 *As old as the Puritans*: Increase Mather, for example, expected the world's destruction by fire—and the survival of the righteous, who would "be *caught up into the Air*" beforehand. See Boyer, *When Time Shall Be No More*, p. 76.

63 *"Then we who are alive"*: 1 Thessalonians 4:17.

63 *"Spiritual body"*: 1 Corinthians 15:44.

64 *Whether here or in the hereafter*: See L. M. White, *From Jesus to Christianity* (San Francisco: HarperSanFrancisco, 2004).

65 *American slaves*: For the definitive account of nineteenth-century slave religion, see A. J. Raboteau, *Slave Religion: The "Invisible Institution" in the Antebellum South* (New York: Oxford University Press, 1978).

65 *"And you'll see de stars"*: Ibid., p. 263.

65 *"In the life to come"*: E. Burke, *Pleasure and Pain: Reminiscences of Georgia in the 1840s* (Savannah: Beehive Press, 1978), quoted in Raboteau, *Slave Religion*, pp. 291–92.

66 *Isaiah promised*: Isaiah 11:1.

66 *Daniel and Enoch promised*: Daniel 12:3; see also H. T. Andrews, *The Apocryphal Books of the Old and New Testament* (London: T. C. & E. C. Jack, 1908), p. 79.

66 *"Neither apostolic nor prophetic"*: M. Luther, "Preface to the Revelation of St. John," in *Luther's Works*, vol. 35, ed. E. T. Bachmann (Philadelphia: Fortress, 1960), pp. 398–99.

66 *Luther's New Testament*: In both the first printing of Luther's New Testament, from September 1522, and the second, from December 1522, Revelation is listed separately from the approved Lutheran canon. Thanks to the librarians at the American Bible Society for showing me this book.

66 *Crucial to our story*: For a discussion of Revelation's broader impact, see J. Kirsch, *A History of the End of the World: How the Most Controversial Book in the Bible Changed the Course of Western Civilization* (San Francisco: HarperSanFrancisco, 2006).

66 *All the images of heaven*: Angels singing are in Revelation 4:8, although see also Isaiah 6:3; God's throne is in Revelation 4:1–11; singing saints are in Revelation 7:9–13; gold streets, jeweled walls, and pearl gates are in Revelation 21:18–21; garden and rivers are in Revelation 22:1–2.

67 *The narrative thread*: Introductions are made in Revelation 1; steeds and warriors are unleashed in 19:11–17; dragons and beasts appear in chapters 12 and 13; the sun blackens in 6:12–13; and the new Jerusalem descends in 21:1–2. Note that the image of stars falling is also found in the Qur'an; see sura 81 ("Rolling Up").

67 *Horror-movie images*: The sharp sword is from Revelation 19:13–15; the slain

lamb is from 1:12 (and see also 5:5–10); the trumpets are from 8:2; the bowls are from 16; and the crowned dragon is from 12:3.

68　*Other numbers*: Myriads of angels are described in Revelation 5:11; two hundred million troops in 9:16; years of witnessing in 11:3; and the number of the beast in 13:18.

68　*William Miller*: M. A. Noll, *A History of Christianity in the United States and Canada* (Grand Rapids, MI: William B. Eerdmans, 1992), p. 193.

68　*Based on his own calculations*: H. Camping, "We Are Almost There!" Family Stations Inc., February 1, 2008, www.familyradio.com/graphical/literature/waat/waat.pdf.

68　*I met Leonard Thompson*: L. L. Thompson, *The Book of Revelation: Apocalypse and Empire* (New York: Oxford University Press, 1990). Also see his translation: L. L. Thompson, "Apocalypse of John: A Poem of Terrible Beauty" (Menasha, WI: Moulting Mantis Library, 2003).

69　*The author of Revelation*: For a more detailed discussion of Revelation and its provenance, see A. Y. Collins, *Crisis and Catharsis: The Power of the Apocalypse* (Philadelphia: Westminster Press, 1984). Thanks to Adela Collins, Buckingham Professor of New Testament Criticism and Interpretation at Yale Divinity School, for her thoughts on the identity of the author of Revelation and the social circumstances that produced the book.

70　*This ritual made the emperor equivalent*: S. J. Friesen, *Imperial Cults and the Apocalypse of John: Reading Revelation in the Ruins* (Oxford: Oxford University Press, 2001). I am indebted to Steve Friesen, Louise Farmer Boyer Chair in Biblical Studies at the University of Texas at Austin, for his insights.

70　*Capitulation to Rome*: On the question of sacrificial meat, Paul takes a more conciliatory position: "Eat whatever is sold in the meat market without raising any question on the grounds of conscience. . . . If I partake with thankfulness, why should I be denounced because of that for which I give thanks?" (1 Corinthians 10:18–33).

70　*"Those who worship the beast"*: Revelation 14:9–10.

72　*"Light into the world"*: John 12:46.

72　*Origen . . . denied a physical resurrection*: As Caroline Walker Bynum notes, there's room for disagreement about the specifics of Origen's theory of resurrection, not least because his primary treatise on the subject has been lost. Bynum suggests that Origen believed that in heaven we will possess a spiritual body, and that the specific flesh we inhabit in this life will be discarded. C. W. Bynum, *The Resurrection of the Body in Western Christianity, 200–1336* (New York: Columbia University Press, 1995), pp. 63–66. For an alternative reading, see N. T. Wright, *The Resurrection of the Son of God* (Minneapolis: Fortress Press, 2003), pp. 519–21.

72 *"The Kingdom . . . does not come"*: After Luke 17:20–21. Origen, *Origen on Prayer* (Grand Rapids, MI: Christian Classics Ethereal Library, 2001), p. 42.

72 *Heard these words*: Matthew 19:21 and 6:34.

73 *Anthony went out into the desert*: Athanasius, "Life of Anthony," in *A Select Library of Nicene and Post-Nicene Fathers of the Christian Church*, ed. P. Schaff and H. Wace (New York: Charles Scribner's Sons, 1903), vol. 4, pp. 188–221. See also P. R. L. Brown, *The Body and Society: Men, Women, and Sexual Renunciation in Early Christianity* (New York: Columbia University Press, 1988), pp. 213–40.

73 G. Frank, *The Memory of the Eyes: Pilgrims to Living Saints in Christian Late Antiquity* (Berkeley: University of California Press, 2000). Georgia Frank, Professor of Religion at Colgate University, spoke to me eloquently about the connection of the desert landscape to heaven.

73 *"All my senses are overwhelmed"*: M. Gruber, *Journey Back to Eden: My Life and Times Among the Desert Fathers* (Maryknoll, NY: Orbis Books, 2004), p. 20. See also Frank, *The Memory of the Eyes.*

73 *"Pray without ceasing*: 1 Thessalonians 5:17.

73 *"Neither fat . . . nor lean"*: Athanasius, "Life of Anthony."

74 *First Christian hermit*: Although Saint Paul of Thebes also has a strong claim to the title.

74 *Feats of self-abnegation*: J. Chryssavgis, *In the Heart of the Desert: The Spirituality of the Desert Fathers and Mothers* (Bloomington, IN: World Wisdom, 2003); see also H. Thurston, "Simeon Stylites the Elder," in *The Catholic Encyclopedia*, ed. Herbermann et al., vol. 13, p. 795.

74 *Anthony's compatriot Pachomius*: Sozomenus, "The Ecclesiastical History of Sozomen, Comprising a History of the Church from AD 323 to AD 425," in *A Select Library of Nicene and Post-Nicene Fathers of the Christian Church*, 2nd series, ed. P. Schaff and H. Wace (New York: The Christian Literature Company, 1890), vol. 2, p. 292.

74 *Books of aphorisms*: See, for example, T. Merton, *The Wisdom of the Desert* (New York: New Direction Books, 1960). Also B. Ward, *The Desert Fathers: Sayings of the Early Christian Monks* (London: Penguin Books, 2003).

74 *"The present age is a storm"*: L. Swan, *The Forgotten Desert Mothers* (Mahwah, NJ: Paulist Press, 2001), p. 64.

74 *"Let him rest"*: D. Burton-Christie, *The Word in the Desert* (New York: Oxford University Press, 1993), p. 289. My gratitude to Douglas Burton-Christie, professor of theological studies at Loyola Marymount University, for his analysis of heaven and the Desert Fathers.

75 *"Jesus knew"*: J. D. Salinger, *Franny and Zooey* (Boston: Little, Brown, 1966), p. 171.

75   *A conscious effort to replicate heaven*: R. B. Lockhart, *Halfway to Heaven: The Hidden Life of the Sublime Carthusians* (New York: Vanguard Press, 1985).

75   *Reformers continued to debate*: P. Althaus, *The Theology of Martin Luther* (Philadelphia: Fortress Press, 1966), p. 414; see also B. Cottret, *Calvin: A Biography* (Grand Rapids, MI: William B. Eerdmans, 1995), pp. 79–81.

76   *"Go directly to Christ"*: A. C. Cochrane, ed., *Reformed Confessions of the 16th Century* (Louisville, KY: Westminster John Knox Press, 2003), p. 295.

76   *Founded by Christans seeking a home*: M. Holloway, *Heavens on Earth: Utopian Communities in America, 1680–1880* (New York: Dover, 1966).

76   *"Now the whole group"*: Acts 4:32.

76   *"The true END of Life"*: C. Mather, *Manuductio ad Ministerium: Directions for a Candidate of the Ministry* (New York: Columbia University Press, 1938), p. 6.

76   *"As a city upon a Hill"*: O. Collins, ed. *Speeches That Changed the World* (Louisville, KY: Westminster John Knox Press, 1999), pp. 63–65.

76   *The Ephrata cloister*: For more detail on both the Ephrata cloister and the Oneida community, see R. P. Sutton, *Communal Utopias and the American Experience: Religious Communities 1732–2000* (Westport, CT: Praeger, 2003) and Holloway, *Heavens on Earth*.

77   *"Peculiar grace"*: T. Merton, "Introduction," in *Religion in Wood: Masterpieces of Shaker Furniture*, ed. E. D. Andrews and F. Andrews (Bloomington: Indiana University Press, 1966), pp. 7–18.

77   *"Not a reading man"*: R. W. Emerson, "Concord, October 30, 1840," in *The Correspondence of Thomas Carlyle and Ralph Waldo Emerson, 1834–1872* (Cambridge, MA: Riverside Press, 1896), vol. 1, p. 227.

78   *"the perfectibility of human society . . ."*: R. Wuthnow, ed., *Encyclopedia of Politics and Religion*, vol. 2 (Washington, DC: CQ Press, 2007), pp. 829–30.

78   *"A demonstration plot"*: T. E. K'Meyer, *Interracialism and Christian Community in the Postwar South: The Story of Koinonia Farm* (Charlottesville: University Press of Virginia, 2000).

79   *"Radical, into the social gospel"*: Koinonia Partners, "Con Browne's Talks in Late February 2006, as Recorded by Ann Karp," February, 2006, http://www.koinoniapartners.org/History/oralhistory/Con_Browne.html.

79   *"I made it through my trials"*: Koinonia Partners, "Koinonia Memories from Ms. Georgia Solomon as Told to Ann Karp," 2006, www.koinoniapartners.org/History/oralhistory/Georgia_Solomon.html.

CHAPTER FOUR: GREEN, GREEN PASTURES

82   *Muslims in America, most of them immigrants*: For further details, see The Pew Research Center, "Muslim Americans: Middle Class and Mostly Mainstream,"

May 22, 2007, http://people-press.org/report/?reportid=329; and Y. Y. Haddad and J. I. Smith, *Muslim Communities in North America* (Albany: State University of New York Press, 1994), p. 275.

82 *"Know that the gardens"*: B. Woodward, "In Hijacker's Bags, a Call to Planning, Prayer and Death," *Washington Post*, September 28, 2001.

82 *The company of beautiful maidens*: See suras 44 ("Smoke"), 52 ("The Mountain"), 55 ("The All-Merciful"), and 56 ("The Calamity"). For a more complete discussion of the concept's hadithic and Qur'anic roots, see J. I. Smith and Y. Y. Haddad, *The Islamic Understanding of Death and Resurrection* (Oxford: Oxford University Press, 2002), pp. 163–68.

83 *"Eyes large and dark"*: Sura 56 ("The Calamity"), verse 23.

83 *"White raisin"*: A. Stille, "Scholars Are Quietly Offering New Theories of the Koran," *New York Times*, March 2, 2002.

83 *During the Iran-Iraq war*: L. Miller, "Why We Need Heaven," *Newsweek*, August 12, 2002.

83 *Sublime satisfactions*: See, for example, M. K. Nydell, *Understanding Arabs: A Guide for Modern Times* (Yarmouth, ME: Intercultural Press, 2006), p. 109.

83 *"The clutches of religion"*: J. R. Petersen, "Virgins in Paradise: The Strange Erotic Visions of a Suicide Bomber," *Playboy*, April 2002.

85 *"All they do is noted"*: Sura 54 ("The Moon"), verses 52–53. I've quoted here from the Yusuf Ali translation: A. Yusuf Ali, *The Qur'an: Translation* (New York: Tahrike Tarsile Qur'an, 2007). Khalidi translates the lines as "All they have done is in ancient Scriptures / And all of it, great or small, is recorded."

85 *A book, or a scroll*: Smith and Haddad, *Islamic Understanding of Death*, p. 76. See also Sura 84 ("The Splitting").

85 *"They whose scales"*: Sura 23 ("The Believers"), verses 102–4.

86 *"Scorching wind"*: Sura 56 ("The Calamity"), verses 40–43.

86 *"God is up there"*: Associated Press, "Diamonds Can't Tempt Honest LA Cabbie," MSNBC, November 18, 2005, http://www.msnbc.msn.com/id/10096893/.

86 *"The sun, entering as a tyrant"*: C. M. Doughty, *Travels in Arabia Deserta* (Cambridge: Cambridge University Press, 1888), p. 323.

87 *Parents sometimes buried newborn girls*: See, for example, Sura 16 ("The Bee"), verses 58–59: "Yet when one of them is brought tidings of an infant girl, his face turns dark, suppressing his vexation. He keeps out of people's sight, because of the evil news he was greeted with. Will he retain the infant, in disgrace, or will he bury it in haste in the ground? Wretched indeed is their decision!"

87 *Bedouin religious practices*: Some experts warn, however, against judging rural, small-scale religious practices from an urban—and literally "civilized"—perspective. For a more detailed discussion of Bedouin religious practices in this

period, see J. Henninger, "Pre-Islamic Bedouin Religion," in *Studies on Islam*, ed. M. L. Swartz (New York: Oxford University Press, 1981), pp. 3–22.

87 *They did have religious traditions*: S. Inayatullah, "Pre-Islamic Thought," in *A History of Muslim Philosophy*, ed. M. M. Sharif (Kempten, Pakistan: Philosophical Congress, 1963), vol. 1, pp. 126–35.

88 *"I will cry for you"*: M. Abdesselem, *Le thème de la mort dans la poésie arabe des origines à la fin du IIIe/IXe siècle* (Tunis, Tunisia: Publications de l'Université de Tunis, 1977), p. 63. Translation courtesy of Ana Rodriguez Navas, Department of Comparative Literature, Princeton University.

88 *"Why does death persecute us"*: Abdesselem, *Le thème de la mort*, p. 69. Translation courtesy of Ana Rodriguez Navas, Department of Comparative Literature, Princeton University.

88 *"Suffocating heat"*: W. M. Watt, A. J. Wensinck, R. B. Winder, and D. A. King, "Makka," in *Encyclopaedia of Islam, Second Edition*, ed. P. Bearman, T. Bianquis, C. E. Bosworth, E. v. Donzel, and W. P. Heinrichs (Leiden: E. J. Brill, 2009), vol. 6, p. 144. See also F. E. Peters, *Mecca: A Literary History of the Muslim Holy Land* (Princeton, NJ: Princeton University Press, 1994).

89 *The great British scholar*: See, for instance, W. M. Watt, *Muhammad at Mecca* (Oxford: Clarendon Press, 1953); and W. M. Watt, *Muhammad at Medina* (Oxford: Clarendon Press, 1956). For more recent biographies, see M. Rodinson, *Muhammad* (London: Tauris Parke, 2002); and K. Armstrong, *Muhammad: A Biography of the Prophet* (San Francisco: HarperSanFrancisco, 1993).

89 *No record keepers*: For a broader discussion of literacy in the pre-Islamic Hijaz, and the resultant historiographical challenges, see C. F. Robinson, *Islamic Historiography* (Cambridge: Cambridge University Press, 2003), pp. 3–17. For a more general overview of Islamic history, see B. Lewis and B. E. Churchill, *Islam: The Religion and the People* (Upper Saddle River, NJ: Wharton School Publishing, 2008).

90 *Exists exactly as it is*: See Sura 43 ("Ornament"), verses 1–4. The concept of a heavenly original of the Qur'an is also found in the hadiths.

90 *A new generation of scholars*: See especially P. Crone, *Meccan Trade and the Rise of Islam* (Princeton, NJ: Princeton University Press, 1987); and P. Crone, *From Arabian Tribes to Islamic Empire: Army, State and Society in the Near East c. 600–850* (Aldershot, UK: Ashgate, 2008).

90 *Also under renovation*: F. M. Donner, "The Background to Islam," in *The Cambridge Companion to the Age of Justinian*, ed. M. I. Maas (New York: Cambridge University Press, 2005), pp. 510–34.

91 *Muhammad placed his hands*: D. N. Freedman and M. J. McClymond, *The Rivers of Paradise: Moses, Buddha, Confucius, Jesus, and Muhammad as Religious Founders* (Grand Rapids, MI: William B. Eerdmans, 2001), p. 584.

91 *The Qaynuqa*: Watt, *Muhammad at Medina*, p. 194.

91 *Muhammad was born*: W. M. Watt, *Muhammad: Prophet and Statesman* (London: Oxford University Press, 1974), p. 7.

91 *Sensitive, introspective, and spiritual*: See, for example, the hagiography of twentieth-century Egyptian politician and author Muhammad Husayn Haykal, who praises the twelve-year-old Muhammad's "largeness of spirit, intelligence of heart, superior understanding . . . accurate memory, and other characteristics which the divine providence gave him in preparation for his great mission." Translated in A. Wessels, *A Modern Arabic Biography of Muhammad* (Leiden: E. J. Brill, 1972), p. 50.

91 *A formative experience*: As Crone notes in *Meccan Trade*, the Islamic tradition abounds with tales of the young Muhammad meeting representatives of non-Islamic religions and being acknowledged as a future prophet; in some versions of this story, Bahira is a Jewish rabbi.

92 *Kadijah . . . could afford*: Watt, *Muhammad at Mecca*.

92 *In the year 610*: Ibid. Watt notes that some early authorities put the date of Muhammad's prophethood a few years later, around 613 CE.

92 *"Recite!"*: Watt, in *Muhammad: Prophet and Statesman*, notes that Muhammad originally believed the angelic presence to be God himself. For an account translated directly from the pertinent hadiths, see M. Lings, *Muhammad: His Life Based on the Earliest Sources* (London: Islamic Texts Society, 1983), p. 43.

92 *Physically squeezed him*: See, for example, K. Armstrong, *Islam: A Short History* (New York: Modern Library, 2002), p. 4.

92 *"Recite, in the name of your Lord!"*: Sura 96 ("The Blood Clot").

93 *Islam's "five pillars"*: The five pillars are perhaps most clearly expressed in the "Hadith of the Angel Gabriel." See V. J. Cornell, *Voices of Islam* (Westport, CT: Praeger, 2007), pp. 7–8.

93 *"Simply not working"*: Armstrong, *Islam: A Short History*, p. 8.

93 *The paradise described in the Qur'an*: Green pastures are described in Sura 40 ("The All-Merciful"), verses 60–75; I have followed the translation given in A. J. Arberry, *The Koran Interpreted* (New York: Touchstone, 1996). For rivers of water and milk, see Sura 47 ("Muhammad"), verses 12–15; for pomegranates, see Sura 55 ("The All-Merciful"), verse 68; for "flesh of fowl," see Sura 56 ("The Calamity"), verse 21; for robes and goblets, see Sura 76 ("Man"); and for heavenly greetings, see Sura 14 ("Ibrahim"), verse 23.

94 *"Do you speak Arabic?"*: I am grateful to Jonathan Brockopp for this anecdote.

94 *A day will come*: Sura 81 ("Rolling Up").

95 *Up to their necks in sweat*: Smith and Haddad, *Islamic Understanding of Death*, pp. 73–76, See also L. Halevi, *Muhammad's Grave: Death Rites and the Making of Islamic Society* (New York: Columbia University Press, 2007).

95 *Judgment is handed down*: See sura 23 ("The Believers"), verses 102–4, and Sura 84 ("The Splitting").

95 *A bridge*: Smith and Haddad, *Islamic Understanding of Death*, pp. 78–80.

95 *"Garments of fire"*: Sura 22 ("The Pilgrimage"), verse 18–20.

95 *"We shall not be resurrected"*: Sura 23 ("The Believers"), verse 37.

95 *"If you are in doubt"*: Sura 22 ("The Pilgrimage"), verses 4–6.

96 *The great man at the center*: P. Barrett, *American Islam: The Struggle for the Soul of a Religion* (New York: Farrar, Straus and Giroux, 2007), p. 63.

97 *"Produced a culture"*: K. Abou El Fadl, "What Became of Tolerance in Islam?" *Los Angeles Times*, September 14, 2001. See also K. Abou El Fadl, *The Great Theft: Wrestling Islam from the Extremists* (New York: HarperCollins, 2005).

98 *Received death threats*: Barrett, *American Islam*, pp. 92–93.

98 *What the Qur'an means*: "Throne," Surah 11 ("Hud"), verse 7; "jugular vein," Sura 50 ("Qaf"), verse 16; "light of the heavens," Sura 24 ("Light").

99 *Barrett . . . visits Abou El Fadl*: Barrett, *American Islam*, pp. 67, 93.

100 *"A preacher was saying one day"*: G. Morrison, ed., *History of Persian Literature from the Beginning of the Islamic Period to the Present Day* (Leiden: E. J. Brill, 1981), p. 65.

101 *"Banish . . . and beat them"*: Arberry, *The Koran Interpreted*, pp. 105–6. For a discussion of Bakhtiar's translation, and other possible renderings of the verse, see A. Saeed, *The Qur'an: An Introduction* (New York: Routledge, 2008), pp. 129–33.

101 *"Abandon them in their sleeping places"*: L. Bakhtiar, *The Sublime Quran* (Chicago: Kazi Publications, 2007), p. 70. See also her introductory essay, pp. xxvii–xxxiv.

101 *Upheld Bakhtiar's translation*: I. Mattson, "RE: Statements Made by ISNA Canada Secretary General Regarding Dr. Laleh Bakhtiar's Qur'an Translation," The Islamic Society of North America, October 24, 2007, http://www.isna.net/articles/Press-Releases/PUBLIC-STATEMENT.aspx.

102 *"Lovely eyed ones"*: Bakhtiar, *The Sublime Quran*, Sura 56 ("The Calamity"), verse 23.

CHAPTER FIVE: RESURRECTION

106 *"So I believe in an absurdity"*: C. Milosz, *To Begin Where I Am* (New York: Farrar, Straus and Giroux, 2001), p. 320.

107 *Only 26 percent*: D. Van Biema, "Does Heaven Exist?" *Time*, March 24, 1997.

107 *2003 Harris poll*: "The Religious and Other Beliefs of Americans 2003," Harris Interactive, February 26, 2003, http://www.harrisinteractive.com/harris_poll/index.asp?pid=359.

107 *Resurrection belief is fading*: Cremation Association of North America, "Final 2005 Statistics and Projections to the Year 2025; 2006 Preliminary Data," 89th Annual Convention of the Cremation Association of North America, San Francisco, August 15–18, 2007. See also S. Prothero, *Purified by Fire: A History of Cremation in America* (Berkeley, CA: University of California Press, 2001).

108 *Recently changed its position*: J. Newton, "Catholic Church (North America)," in *The Encyclopedia of Cremation*, ed. D. J. Davies and L. H. Mates (Burlington, VT: Ashgate, 2005), p. 112. Note, however, that some groups—Orthodox Jews, Eastern Orthodox Christians—do still continue to resist cremation.

108 *"Formed man from the dust"*: Genesis 2:7.

109 *"Yes indeed!"*: Sura 75 ("Resurrection"), verses 1–5.

110 *"Whole, cognizant, and responsible"*: J. I. Smith and Y. Y. Haddad, *The Islamic Understanding of Death and Resurrection* (Oxford: Oxford University Press, 2002), p. 64.

110 *"I believe with perfect faith"*: J. D. Bleich, ed., *With Perfect Faith: The Foundations of Jewish Belief* (New York: Ktav, 1982), p. 13; see also pp. 638–56.

110 *A traditional prohibition*: A. Steinberg and F. Rosner, *Encyclopedia of Jewish Medical Ethics: A Compilation of Jewish Medical Law on All Topics of Medical Interest* (Jerusalem: Feldheim, 2003), p. 1097.

110 *With soil from Israel*: See, for example, Holy Land Earth, "Suggested Uses for Holy Land Earth," Holy Land Earth LLC, 2009, www.holylandearth.com/uses.asp.

111 *Scholars fight bitterly*: For a discussion of the academic debate surrounding the date and location of Ezekiel's composition, see G. W. Bromiley, *International Standard Bible Encyclopedia* (Grand Rapids, MI: William B. Eerdmans, 1979), pp. 250–62.

111 *Book of Ezekiel*: "Can these bones," Ezekiel 37:3; "a vast multitude," Ezekiel 37:10; "the whole house of Israel," Ezekiel 37:11–14.

112 *Levenson is a man on a mission*: See also K. Madigan and J. D. Levenson, *Resurrection: The Power of God for Christians and Jews* (New Haven, CT: Yale University Press, 2008).

112 *Restoration, purification, and renewal*: J. D. Levenson, *Resurrection and the Restoration of Israel: The Ultimate Victory of the God of Life* (New Haven, CT: Yale University Press, 2006).

113 *Say the Amidah*: L. I. Levine, *The Ancient Synagogue: The First Thousand Years* (New Haven, CT: Yale University Press, 2005), pp. 540–50.

113 *"Immortality of the soul"*: M. A. Meyer, *Response to Modernity: A History of the Reform Movement in Judaism* (New York: Oxford University Press, 1988), pp. 228–29.

113 *"Entirely foreign . . . not rooted in Judaism"*: Central Conference of American

Rabbis, "Pittsburgh Platform: Declaration of Principles," in *Religion and American Cultures*, ed. G. Laderman and L. D. León (Santa Barbara, CA: ABC-CLIO, 2003), p. 779.

114   *Reinserted the ancient avowal*: B. Harris, "Reform Siddur Revives Resurrection Prayer," JTA (Jewish Telegraph Agency), November 20, 2007.

115   *The body was a trap*: Plato, *Phaedo* (Stillwell, KS: Digireads.com, 2006), p. 58.

115   *"Under the inspiration of God"*: See chapter 6 of Clement of Alexandria's *Exhortation to the Heathen*, quoted in B. M. Metzger, *The Canon of the New Testament: Its Origin, Development, and Significance* (Oxford: Clarendon Press, 1997), p. 134.

115   *"Among the pupils of Socrates"*: Augustine, *The City of God Against the Pagans* (Cambridge: Cambridge University Press, 1998), p. 316.

115   *The Pharisees*: J. H. Charlesworth and C. D. Elledge, *Resurrection: The Origin and Future of a Biblical Doctrine* (New York: T. & T. Clark, 2006), pp. 36–41.

116   *"You know neither the scriptures"*: Mark 12:18–25.

116   *A priest or a king*: 1 Chronicles 17:11–15.

116   *Not from the house of David*: While Matthew (in Matthew 1:1–17) goes out of his way to present Jesus as a descendant of David, many scholars believe that the genealogy he offers is aspirational rather than strictly historical. See R. H. Williams, "An Illustration of Historical Inquiry: Histories of Jesus and Matthew 1:1–25," in *Handbook of Early Christianity*, ed. A. J. Blasi, P.-A. Turcotte, and J. Duhaime (Walnut Creek, CA: AltaMira Press, 2002), pp. 105–24.

116   *The task of the Gospel writers*: See C. W. Bynum, *The Resurrection of the Body in Western Christianity, 200–1336* (New York: Columbia University Press, 1995), pp. 4–5.

116   *Each of the Gospels*: See Matthew 28:9; Mark 16:9–15; Luke 24:15–31; and John 20:27. Note that some scholars believe Mark's gospel originally ended with Mark 16:8 and that all subsequent verses were added by early Christians seeking to bring the book to a more satisfying conclusion. See chapter 2 ("Desire for an End," pp. 34–56) in G. Aichele, *Jesus Framed* (London: Routledge, 1996).

117   *A particularly vigorous defense*: 1 Corinthians 15:35–55. See also Bynum, *The Resurrection of the Body*, which begins with the "seed" image and explores its themes throughout.

120   *"Mouths will no longer eat"*: Bynum, *The Resurrection of the Body*, pp. 37–42.

120   *Almost single-handedly*: I owe a great debt of gratitude to Paula Fredriksen who, in many phone calls and e-mails, was my guide to Augustine and his world. See also P. Fredriksen, *Augustine and the Jews: A Christian Defense of Jews and Judaism* (New York: Doubleday, 2008); G. Wills, *Saint Augustine* (New York: Lipper/Viking, 1999); and A. Fitzgerald and J. C. Cavadini, *Augustine Through the Ages: An Encyclopedia* (Grand Rapids, MI: William B. Eerdmans, 1999).

120 *The son of a Christian mother:* The best biography is still P. R. L. Brown, *Augustine of Hippo: A Biography* (Berkeley: University of California Press, 2000); see also P. R. L. Brown, *The Body and Society* (New York: Columbia University Press, 1988).

121 *Among Manicheans:* R. M. Hogan, *Dissent from the Creed: Heresies Past and Present* (Huntington, IN: Our Sunday Visitor, 2001), pp. 61–66.

121 *"Flight and escape":* Augustine, "The Magnitude of the Soul," trans. J. J. McMahon, in *The Fathers of the Church: A New Translation,* vol. 4, ed. R. J. Deferrari (Washington, DC: Catholic University of America Press, 1947), p. 144.

121 *"Squared the circle":* See, especially, P. Frederiksen, "Vile Bodies: Paul and Augustine on the Resurrection of the Flesh," in *Biblical Hermeneutics in Historical Perspective: Studies in Honor of Karlfried Froehlich on His Sixtieth Birthday,* ed. M. S. Burrows and P. Rorem (Grand Rapids, MI: William B. Eerdmans, 1991), pp. 75–87.

122 *"Devoured by beasts":* Book 22, chapter 12, in Augustine, "The City of God Against the Pagans," in *A Select Library of the Nicene and Post-Nicene Fathers of the Christian Church,* ed. P. Schaff (New York: Charles Scribner's Sons, 1899), vol. 2, p. 494.

122 *The rationalistic questions:* Ibid. People rise aged 30, book 22, chapter 15; marks of martyrdom, book 22, chapter 19; babies rise as adults, book 22, chapter 14; women rise as women, book 22, chapter 17.

122 *Unbaptized babies:* Augustine, "On Merit and the Forgiveness of Sins, and the Baptism of Infants," *New Advent,* 2008, http://www.newadvent.org/fathers/15011.htm.

122 *"The body is made incorruptible":* Book 22, chapter 30, in Augustine, *City of God,* p. 1178.

123 *"We shall be still and see":* Book 22, chapter 30, in Augustine, *City of God,* trans. H. Bettenson (London: Penguin, 2003), p. 1082.

CHAPTER SIX: SALVATION

125 *Obama . . . convened a meeting:* Associated Press, "Obama Reaches Out to Christian Leaders," MSNBC, June 11, 2008, http://www.msnbc.msn.com/id/25092483/.

125 *"The only way for me":* A. M. Banks and D. Burke, "Fuller Picture Emerges of Obama's Evangelical Meeting," Religion News Service, June 18, 2008, http://pewforum.org/news/display.php?NewsID=15867.

126 *"I do not believe she went to hell":* L. Miller and R. Wolffe, "Finding His Faith," *Newsweek,* July 12, 2008.

126 *"I am a big believer"*: L. Miller and R. Wolffe, "I Am a Big Believer in Not Just Words, But Deeds and Works," *Newsweek*, July 21, 2008.

126 *"My particular set of beliefs"*: Miller and Wolffe, "Finding His Faith."

126 *Works versus Grace*: For a discussion of the grace-versus-works debate in Judaism, see K. L. Yinger, *Paul, Judaism, and Judgment According to Deeds* (Cambridge: Cambridge University Press, 1999). While Islam is sometimes portrayed as entirely works-focused, many Muslim theologians argue that good deeds and submission to God are themselves contingent upon God's grace; see, for example, A. Schimmel, "Some Aspects of Mystical Prayer in Islam," *Die Welt des Islams*, New Series, vol. 2, no. 2 (1952): 112–25.

126 *Can have devastating consequences*: Thanks in particular to Jerry Walls, of Asbury Theological Seminary, for his insights on Salvation theory. See J. L. Walls, *Heaven: The Logic of Eternal Joy* (New York: Oxford University Press, 2007).

127 *"About two minutes ago"*: For other examples of "pearly gate" jokes, see D. Capps, *A Time to Laugh: The Religion of Humor* (New York: Continuum, 2005), pp. 30–33.

127 *"Your place is in heaven"*: M. Wagner, "Shas Pulls Controversial Election Ad," *The Jerusalem Post*, March 12, 2006.

128 *"We have been waiting"*: R. Ashmore, "McVeigh Will Meet Unimaginable Mercy," *National Catholic Reporter*, May 18, 2001.

128 *"But they will all go to hell"*: "Perspectives," *Newsweek*, August 24 and 31, 2009.

128 *"This store burns souls"*: A. Teibel, "Embracing Secular Culture Can Be Risky in Israel," Associated Press, October 5, 2008.

129 *According to a 2008 poll*: The Pew Forum on Religion and Public Life, *U.S. Religious Landscape Survey: Religious Beliefs and Practices*, Washington, DC, 2008.

129 *"Saving faith"*: Southern Baptist Theological Seminary, "What We Believe," 2009, http://www.sebts.edu/about/what-we-believe/default.aspx.

129 *Pew promptly polled directly*: The Pew Forum on Religion and Public Life, "Many Americans Say Other Faiths Can Lead to Eternal Life," December 18, 2008, http://pewforum.org/docs/?DocID=380.

129 *"Decisions only the Lord will make"*: J. Meacham, "Pilgrim's Progress," *Newsweek*, August 14, 2006.

130 *Denounced as an apostate*: T. Flannery, "Billy Graham's Apostasy," WorldNetDaily, August 10, 2006, http://www.worldnetdaily.com/news/article.asp?ARTICLE_ID=51461.

130 *"Do as you will"*: Zwingli's "Plague Song" is included in E. Egli and G. Finsler (eds.), *Huldreich Zwinglis sämtliche Werke*, Volume 1 (Berlin: C. A. Schwetschke und Sohn, 1905), p. 67. I have followed the translation given in A. E. McGrath, *Reformation Thought: An Introduction* (Oxford: Blackwell, 1999), p. 133.

130 *Predestination*: The concept of predestination existed in Pauline literature, but it was Augustine, in his letters against the Pelagians (a group deemed heretical

that emphasized free will), who first popularized it. P. R. L. Brown, *Augustine of Hippo: A Biography* (Berkeley: University of California Press, 2000).

130 *An angel . . . sprinkles soil*: J. I. Smith and Y. Y. Haddad, *The Islamic Understanding of Death and Resurrection* (Oxford: Oxford University Press, 2002), pp. 35–36.

131 *"Who shall live"*: R. Hammer, *Entering the High Holy Days: A Guide to Origins, Themes, and Prayers* (Philadelphia: Jewish Publication Society of America, 2005), p. 87.

131 *"Lead many to righteousness"*: Daniel 12:3.

131 *"Righteous" Christians*: Revelation 2:14–20.

131 *"The earth shall be inherited"*: Sura 21 ("The Prophets"), verse 105.

132 *"Smiling in their agonies"*: Josephus, "The Jewish War," in *Jewish Life and Thought Among Greeks and Romans*, ed. L. H. Feldman and M. Reinhold (Minneapolis: Augsburg Fortress, 1996), pp. 248–51.

132 *Blood sacrifice*: B. Lang, *Sacred Games: A History of Christian Worship* (New Haven, CT: Yale University Press, 1997), p. 222.

132 *Salvation of all*: Thanks to Craig Townsend, vicar of St. James Church in New York City, for his careful eye and his familiarity with Christian liturgy.

133 *"Let fire and the cross"*: A. Roberts and J. Donaldson, eds., *The Writings of the Apostolic Fathers* (London: Elibron Classics, 2005), p. 214.

133 *"The seed of the church"*: O. Chadwick, *A History of Christianity* (London: Weidenfeld & Nicolson, 1995), p. 35.

133 *"This is my body"*: Matthew 26:26–29. Almost the same words occur in Mark and Luke; the earliest version of the scene can be found in 1 Corinthians 11:23–26.

134 *"Yearn for everlasting life"*: Benedict, *Rule of Saint Benedict in English* (Collegeville, MN: Liturgical Press, 1981), p. 28.

135 *"Clothe not yourself"*: S. Israel, *Charge!: History's Greatest Military Speeches* (Annapolis, MD: Naval Institute Press, 2007), p. 49.

135 *Special favor in heaven*: Perhaps to mitigate their austere earthly existence, twelfth- and thirteenth-century Cistercians actively promoted the idea that they would receive special treatment in the afterlife. In some tellings, deceased monks went first to a gentle, rather attractive version of purgatory, similar to a cloister, where they would be visited regularly by the Virgin Mary; then graduated to a blissful paradise set aside only for members of their order. B. P. McGuire, "A Lost Clairvaux Exemplum Found: The Liber Visionum Et Miraculorum Compiled Under Prior John of Clairvaux," *Analecta Cisterciensia* 39, no. 1 (January–June 1983): 27–62. See also the *Treatise on the Purgatory of St. Patrick*, detailed in C. Zaleski, *Otherworld Journeys: Accounts of Near-Death Experience in Medieval and Modern Times* (New York: Oxford University Press, 1987), p. 38.

136 *Auxiliary prayers for their dead*: C. Daniell, *Death and Burial in Medieval England, 1066–1550* (London: Routledge, 1997), p. 179.

138 *Eight deadly sins*: Evagrius, "On the Eight Thoughts," in *Evagrius of Pontus: The Greek Ascetic Corpus*, ed. R. E. Sinkewicz (Oxford: Oxford University Press, 2006), pp. 66–90.

138 *Edited down to the familiar seven*: Gregory, *Morals on the Book of Job* (Oxford: F. and J. Rivington, 1850), p. 490; see also U. Voll and S. A. Kenel, "Deadly Sins," in *New Catholic Encyclopedia* (Detroit: Thomson/Gale, 2003), vol. 4, pp. 565–67. For more detail on the role of Gregory the Great in the development of a theory of salvation, see J. A. Trumbower, *Rescue for the Dead: The Posthumous Salvation of Non-Christians in Early Christianity*, Oxford Studies in Historical Theology (New York: Oxford University Press, 2001), pp. 141–53.

138 *Penance was public*: J. T. McNeill and H. M. Gamer, *Medieval Handbooks of Penance: A Translation of the Principal "Libri Poenitentiales" and Selections from Related Documents* (New York: Columbia University Press, 1990). P. J. Geary, "Penance (The West)," in *Encyclopedia of the Middle Ages* (e-reference ed.), ed. A. Vauchez, trans. A. Walford (Chicago: Fitzroy Dearborn, 2000), www .oxford-middleages.com/entry?entry=t179.e2155-s1.

138 *Not yet fine-tuned*: Per J. Le Goff, *The Birth of Purgatory* (Chicago: University of Chicago Press, 1984), p. 217; the concept of venial sins as distinct from mortal sins was likely formalized, along with the doctrine of purgatory, around the twelfth century.

138 *"When I am grown old"*: See W. C. Placher, *A History of Christian Theology: An Introduction* (Philadelphia: Westminster Press, 1983), p. 132.

138 *Ledgers of sins*: Thanks to Jeffrey Burton Russell for pointing me to the penitentials.

139 *Not for sissies*: McNeill and Gamer, *Medieval Handbooks of Penance*, pp. 98–116.

139 *Aquinas made distinctions*: I am indebted to Timothy Renick, director of religious studies at Georgia State University, for his guidance on Aquinas. See T. Renick, *Aquinas for Armchair Theologians* (Louisville, KY: Westminster John Knox Press, 2002); and J. P. Wawrykow, *The Westminster Handbook to Thomas Aquinas* (Louisville, KY: Westminster John Knox Press, 2005), pp. 142–45.

139 *"The more love"*: T. Aquinas, *Summa Theologica* 1:12:6. Translation given in C. McDannell and B. Lang, *Heaven: A History* (New York: Vintage Books, 1990), p. 90.

139 *Busy sorting and weighing sins*: S. B. Nuland, *Maimonides* (New York: Nextbook/ Schocken, 2005). See also J. L. Kraemer, *Maimonides: The Life and World of One of Civilization's Greatest Minds* (New York: Doubleday, 2008).

140 *The number 613*: The number 613 is also traditionally taken to represent 365 negative commandments (one for each day of the year) plus 248 positive com-

mandments (one for each organ of the body). See A. Rothkoff, "Mitzvah," in *Encyclopaedia Judaica*, ed. M. Berenbaum and F. Skolnik (Detroit: Macmillan Reference USA, 2007), vol. 14, p. 372.

140 *Thirteen general principles*: H. A. Davidson, *Moses Maimonides: The Man and His Works* (New York: Oxford University Press, 2005), pp. 157–173.

141 *Physician to the sultan*: F. Rosner, *The Medical Legacy of Moses Maimonides* (Hoboken, NJ: Ktav, 1998), p. 14.

143 *Roman Catholic hierarchy*: Le Goff, *The Birth of Purgatory*.

144 *"Neither glorified nor punished"*: Orations 40:23 ("The Oration on Holy Baptism"), in G. Nazianzen, "Select Orations of Saint Gregory Nazianzen," in *A Select Library of Nicene and Post-Nicene Fathers of the Christian Church*, 2nd series, ed. P. Schaff and H. Wace (Grand Rapids, MI: T. & T. Clark), vol. 7, pp. 203–436.

144 *Augustine was harsh*: Le Goff, drawing on the work of Joseph Ntedika, notes that Augustine's early discussions of purgatory were somewhat cursory; it was only after 413, when Augustine squared off against the universalist theories of the *misericordes*, that he began to formulate a more sharply defined theory of post-death purgation as an extension of earthly penance. See Le Goff, *The Birth of Purgatory*, pp. 61–85.

145 *"God . . . adopts our babies"*: "Institutes of the Christian Religion" (1536), quoted in B. A. Gerrish, "The Place of Calvin in Christian Theology," in *The Cambridge Companion to John Calvin*, ed. D. K. McKim (Cambridge: Cambridge University Press, 2004), pp. 289–304.

145 *"Grant me the grace"*: G. Chaucer, *The Canterbury Tales* (Mineola, NY: Dover, 2004), p. 534.

146 *"The third group shall go down"*: Le Goff, *The Birth of Purgatory*, pp. 39–40.

146 *Kaddish for twelve months*: An excellent guide to Kaddish is M. Lamm, *The Jewish Way in Death and Mourning* (Middle Village, NY: Jonathan David Publishers, 2000). See also A. Diamant, *Saying Kaddish: How to Comfort the Dying, Bury the Dead, and Mourn as a Jew* (New York: Schocken, 1998).

146 *Barzakh*: Smith and Haddad, *Islamic Understanding of Death*, pp. 41–52.

147 *"The good go either at once"*: Le Goff, *The Birth of Purgatory*, p. 165.

148 *"The Legend of the Purgatory of St. Patrick"*: See the version in E. Gardiner, ed., *Visions of Heaven and Hell Before Dante* (New York: Italica Press, 1989), pp. 135–48.

148 *"So replete with suffering"*: T. Aquinas, *Summa Theologica*, vol. 5, pt. 3, 2nd sect. and suppl. (New York: Cosimo, 2007), pp. 3006–7.

149 *Popes were granting indulgences*: M. Purcell, *Papal Crusading Policy: The Chief Instruments of Papal Crusading Policy and Crusade to the Holy Land from the Final Loss of Jerusalem to the Fall of Acre 1244–1291* (Leiden: E. J. Brill, 1975),

pp. 36–61. See also D. Webb, *Pilgrims and Pilgrimage in the Medieval West* (London: I. B. Tauris, 1999), pp. 65–66.

149   *"Diriges . . . tyme of my burial"*: E. Duffy, *The Stripping of the Altars: Traditional Religion in England, c. 1400–c. 1580* (New Haven, CT: Yale University Press, 2005), pp. 346, 360–62.

149   *The prayers of the poor*: Ibid., pp. 360–62.

150   *The abuse of indulgences*: P. Collinson, *The Reformation* (London: Weidenfeld & Nicolson, 2003).

150   *Prayer to Saint Gertrude*: Mission to Empty Purgatory, "Calculations," MTEP. com, 2009, http://www.mtep.com/calculations.htm.

150   *Revived the practice of indulgences*: Roman Catholic Diocese of Brooklyn, "Announces Indulgences During Pauline Year," DioceseofBrooklyn.org, August 5, 2008, http://dioceseofbrooklyn.org/default_article.aspx?id=2084. See also J. N. Latino, "Year of St. Paul Offers Indulgences," MississippiCatholic.com, November 14, 2008, http://www.mississippicatholic.com/categories/diocese/2008/111408/indulgences.html.

151   *A test of worthiness*: I am indebted to Kathleen Flake, associate professor of American religious history at Vanderbilt University, for her guidance on Mormonism. For a good overview, see C. L. Bushman and R. L. Bushman, *Building the Kingdom: A History of Mormons in America* (Oxford: Oxford University Press, 2001); and also J. Shipps, *Mormonism: The Story of a New Religious Tradition* (Urbana: University of Illinois Press, 1987).

151   *"Whose glory is that of the sun"*: Doctrine and Covenants 76:70, in "The Official Scriptures of the Church of Jesus Christ of Latter-day Saints," LDS.org, 2006, http://scriptures.lds.org/en/contents. See also Doctrine and Covenants 88:20–39.

151   *A sixth-generation Mormon*: Toscano also appeared on a two-part documentary about Mormonism: *Frontline*, "The Mormons," first broadcast on April 30 and May 1, 2007, by PBS. Produced and directed by H. Whitney, and written by H. Whitney and J. Barnes.

154   *"I lost hold of Christ"*: J. M. Kittelson, *Luther the Reformer: The Story of the Man and His Career* (Minneapolis: Augsburg, 1986), pp. 78–80. See also the first book in Martin Brecht's authoritative three-volume biography: M. Brecht, *Martin Luther: His Road to Reformation 1483–1521* (Minneapolis: Fortress Press, 1993).

154   *"Sobald das Geld"*: The jingle is attributed to Tetzel, perhaps spuriously, by two contemporary witnesses; it's also mentioned by Luther in Theses 27 and 28. P. Schaff, *History of the Christian Church* (New York: Charles Scribner's Sons, 1891), p. 153.

154 *Raped the Virgin*: R. D. Linder, *The Reformation Era* (Westport, CT: Greenwood Press, 2008), p. 22.

154 *"Confident of entering into heaven"*: M. Luther, *Works of Martin Luther, with Introduction and Notes* (Philadelphia: A. J. Holman, 1915), p. 38.

155 *"Equal to St. Paul"*: McDannell and Lang, *Heaven: A History*, p. 150.

CHAPTER SEVEN: VISIONARIES

157 *The story he tells*: D. Piper and C. B. Murphey, *90 Minutes in Heaven: A True Story of Death and Life* (Grand Rapids, MI: Revell, 2004), pp. 25–44. See also D. Piper and C. B. Murphey, *Heaven Is Real: Lessons on Earthly Joy from the Man Who Spent 90 Minutes in Heaven* (New York: Penguin Books, 2007), p. 137.

159 *"Enjoy your life"*: S. Mitchell, *Gilgamesh: A New English Version* (New York: Free Press, 2004), pp. 168–69. See also C. Zaleski, *Otherworld Journeys: Accounts of Near-Death Experience in Medieval and Modern Times* (New York: Oxford University Press, 1987).

159 *Inspired partly by Gilgamesh*: See, for example, W. Burkert, " 'Or Also a Godly Singer': Akkadian and Early Greek Literature," in *Gilgamesh: A Reader*, ed. J. Maier (Wauconda, IL: Bolchazy-Carducci Publishers, 1997), pp. 178–91.

160 *The earliest apocalyptic literature*: For a survey of heavenly apocalyptic literature in Judaism and Christianity, see M. Himmelfarb, *Ascent to Heaven in Jewish and Christian Apocalypses* (Oxford: Oxford University Press, 1993).

160 *"Because God took him"*: Genesis 5:24.

160 *"A wall . . . built of hailstones"*: Zaleski, *Otherworld Journeys*, p. 21.

160 *John . . . is guided to heaven*: "Around the throne" (Revelation, 4:2–5); "a new heaven" (21:1); "Alpha and Omega" (22:13); sent to hell (22:18–21).

161 *"No eye has seen"*: 1 Corinthians 2:9–10. I've followed the New International translation here. *The NIV Study Bible, 10th Anniversary Edition* (Grand Rapids, MI: Zondervan Publishing House, 1995).

161 *"Things that are not to be told"*: 2 Corinthians 12:2–4.

161 *Paul's trip to heaven*: J. K. Elliott, *The Apocryphal New Testament: A Collection of Apocryphal Christian Literature in an English Translation* (Oxford: Oxford University Press, 1993), pp. 624–34. See also E. Gardiner, *Visions of Heaven and Hell Before Dante* (New York: Italica Press, 1988).

162 *"Who carried his servant by night"*: Sura 17 ("The Journey by Night"), verse 1.

162 *Given in the hadith tradition*: See a.-S. M. Ibn 'Alawi, "The Hadith of Isra' and Mi'raj," in *Islamic Doctrines and Beliefs*, vol. 1 (Damascus, Syria: ISCA, 1999), pp. 55–98.

162 *Muhammad and Gabriel*: For the sake of clarity, I've used the familiar spelling, here and elsewhere, for the names of biblical characters. Note, though, that

in most Islamic texts Gabriel is referred to as "Jibril"; Jesus as "Isa"; Enoch as "Idris"; Moses as "Musa"; and so forth.

163 *He sees houris there*: "Miracle of Al-Isra & Al-Miraj," IslamAwareness.net, 2009, http://www.islamawareness.net/Isra/miracle.html.

163 *"The Vision of Tundale"*: All descriptions taken from Gardiner, *Visions of Heaven and Hell*, pp. 149–97.

165 *"His beard is crisped"*: G. Boccaccio and L. B. Aretino, *The Earliest Lives of Dante*, trans. James Robinson Smith (New York: Henry Holt, 1901), pp. 42–43. For a short, pleasurable biography, see P. S. Hawkins, *Dante: A Brief History* (Malden, MA: Blackwell, 2006).

166 *Intended to be news and art*: I am grateful to Peter Hawkins, professor of religion and literature at Yale University's Divinity School and Institute of Sacred Music, for his insights into medieval visionaries and Dante in particular.

166 *Whom Dante allegedly glimpsed*: It's been suggested that "Beatrice" may have been a pseudonym, or even a mere literary device; the abbreviation favored by Dante for the almost-divine object of his affections—"Bice"—could as easily stand for the Latin phrase "Beato Iesu Cristo," or "Blessed Jesus Christ." See J. M. Ferrante, "Beatrice," in Lansing, *The Dante Encyclopedia*, pp. 89–95.

166 *"Nothing else in Western literature"*: H. Bloom, *The Western Canon: The Books and School of the Ages* (New York: Riverhead Trade, 1995), pp. 72–73.

166 *My own volume*: I've used the Hollanders' translation throughout: Dante Alighieri, *Paradiso*, trans. by Robert and Jean Hollander, introduction and notes by Robert Hollander (New York: Doubleday, 2007).

166 *"Rose a living man"*: J. A. Symonds, *The Sonnets of Michael Angelo Buonarroti and Tommaso Campanella* (Teddington, UK: The Echo Library, 2007), p. 14.

167 *Inspired . . . the Sistine Chapel*: For a discussion of the resonances between Dante's epic and Michelangelo's frescoes, see J. L. Miller, *Dante and the Unorthodox: The Aesthetics of Transgression* (Waterloo, ON: Wilfrid Laurier University Press, 2005), pp. 30–32.

167 *All have loved Dante*: See the extensive postscript in N. R. Havely, *Dante* (Malden, MA: Blackwell, 2007).

167 *"Fart and think of Dante"*: D. Bair, *Samuel Beckett: A Biography* (New York: Harcourt Brace Jovanovich, 1978), p. 145.

167 *"As if the prisms"*: S. Heaney, *Station Island* (New York: Farrar, Straus and Giroux, 1985).

168 *"Looked at me with eyes so full"*: Dante, *Paradiso*, canto 4:139–142, p. 91.

168 *Dante's heaven*: "We were in a cloud," Ibid., canto 2:31–36, p. 37; "goodness that is infinite," 33:81, p. 823; "painted with our likeness," 33:131, p. 827; "struck by a bolt," 33:141–42, p. 827; "turning with the Love," 33:143–45, p. 827.

170 *Like a baby*: Ibid., canto 23:121–29, p. 565.

170 *The painter Giotto*: Thanks to Ena Heller, executive director of the Museum of Biblical Art, for pointing me to these pictures.

172 *"Each of the little sparks"*: T. Teeman, "Attention: Towering Intellect at Work," *The Times*, January 24, 2007.

172 *"Do not ask of a vision"*: C. S. Lewis, *The Great Divorce* (New York: Harper-Collins, 2001), p. 144.

173 *"Did our ancestors"*: U. K. Le Guin, "Paradises Lost," in *The Birthday of the World and Other Stories* (New York: HarperCollins, 2002), pp. 213–48.

175 *Mother of modern heaven studies*: I am indebted to Carol Zaleski, professor of world religions at Smith College, for her generous insights. See especially her book *Otherworld Journeys*.

175 *"One second I was alive"*: Piper and Murphey, *Heaven Is Real*, pp. 1–2.

176 *Surprisingly high*: B. Greyson, "Dissociation in People Who Have Near-Death Experiences: Out of Their Bodies or Out of Their Minds?" *The Lancet* 355 (2000): 460–63. Original study published as B. Greyson, "The Incidence of Near-Death Experiences," *Medicine and Psychiatry* 1 (1998): 92–99. Gallup data from 1982 go further, suggesting that up to twenty-three million Americans claim to have had "verge of death" experiences, of whom eight million had "some kind of mystical encounter"; see G. Gallup and W. Proctor, *Adventures in Immortality* (London: Corgi Books, 1984).

176 *Surveys of NDE accounts*: See, for instance, R. A. Moody and P. Perry, *Life Before Life: Regression into Past Lives* (London: Macmillan, 1990). Note, however, that some studies do suggest a decline in the overt religiosity of NDEs since ancient times. In the words of the Dutch scholar Jan Bremmer, modern NDEs "testify to the continuing decline of the afterlife. Heaven is still made of gold and marble, but it is rather empty, except for a few relatives, and even God is no longer there . . . evidently, every age gets the afterlife it deserves." J. N. Bremmer, *The Rise and Fall of the Afterlife: The 1995 Read-Tuckwell Lectures at the University of Bristol* (London: Routledge, 2002), p. 102.

177 *Common across cultures*: J. Belanti, M. Perera, and K. Jagadheesan, "Phenomenology of Near-Death Experiences: A Cross-Cultural Perspective," *Transcultural Psychiatry* 45, no. 1 (2008): 121–33.

177 *Brain scans of religious people*: See, for instance, E. D'Aquili and A. B. Newberg, *The Mystical Mind: Probing the Biology of Religious Experience* (Minneapolis: Augsburg Fortress, 1999). Also A. Newberg and M. R. Waldman, *How God Changes Your Brain: Breakthrough Findings from a Leading Neuroscientist* (New York: Ballantine Books, 2009).

178 *A stiff dose of ketamine*: K. Jansen, "Neuroscience, Ketamine, and the Near-Death Experience: The Role of Glutamate and the NMDA Receptor," in *The*

Near-Death Experience: A Reader, ed. L. W. Bailey and J. Yates (New York: Routledge, 1996), pp. 265–82.

178 *"Into a golden Light"*: K. Jansen, *Ketamine: Dreams and Realities* (Sarasota, FL: The Multidisciplinary Association for Psychedelic Studies, 2000), p. 99.

179 *"There was nothing special"*: C. Dickey, *Summer of Deliverance: A Memoir of Father and Son* (New York: Touchstone, 1998), p. 128.

179 *"Memory cogitating"*: P. Roth, *Indignation* (Boston: Houghton Mifflin, 2008), p. 55.

180 M. Cox-Chapman, *The Case for Heaven: Near-Death Experiences as Evidence of the Afterlife* (New York: G. P. Putnam's Sons, 1995). The book that popularized the near-death experience phenomenon for a generation is R. Moody, *Life After Life* (New York: Bantam, 1978).

181 *Mary Dooley*: Cox-Chapman, *The Case for Heaven*, pp. 14–15.

182 *"The blissful state of the soul"*: M. Maimonides, *Mishneh Torah: The Book of Knowledge* (Jerusalem: Boys Town Jerusalem Publishers, 1965), p. 367.

CHAPTER EIGHT: REUNIONS

184 *Wed in the spirit world*: J. T. Duke, "Marriage: Eternal Marriage," in *Encyclopedia of Mormonism*, ed. D. H. Ludlow (New York: Macmillan, 1992), pp. 857–59.

184 *"Chain of generations"*: B. Young, *Discourses of Brigham Young: Second President of the Church of Jesus Christ of Latter-Day Saints* (South Salt Lake City, Utah: Deseret Book Co., 1971), pp. 406–408.

184 *"Celestial room"*: "Temples: A Virtual Tour of Celestial Rooms," LightPlanet.com, 2008, http://www.lightplanet.com/mormons/temples/celestial_room.html.

184 *Needs to be inducted into heaven*: E. W. Fugal, "Salvation of the Dead," in *Encyclopedia of Mormonism*, ed. Ludlow, pp. 1257–59.

185 *For posthumously baptizing Jews*: G. Niebuhr, "Mormons to End Holocaust Victim Baptism," *New York Times*, April 29, 1995.

186 *1982 Gallup poll*: G. Gallup and W. Proctor, *Adventures in Immortality* (London: Corgi Books, 1984), from C. McDannell and B. Lang, *Heaven: A History* (New York: Vintage Books, 1990), p. 307.

186 *"He and his bride"*: N. R. Kleinfield, "Doors Closed, Kennedys Offer Their Farewells," *New York Times*, July 24, 1999.

186 *"Heaven isn't heaven"*: *The Simpsons*, "Thank God It's Doomsday!" first broadcast on May 8, 2005, on Fox. Directed by M. Marcantel.

186 *"I want to lay my head"*: B. Brown Taylor, "Leaving Myself Behind," in *Heaven*, ed. R. Ferlo (New York: Seabury Books, 2007), p. 10.

187 *"A great homecoming"*: N. R. Gibbs and M. Duffy, *The Preacher and the Presidents: Billy Graham in the White House* (New York: Center Street, 2007), p. 155.

188 *Widespread only in the last two centuries*: For a more detailed account of the development of the family-reunion-in-heaven meme, see McDannell and Lang, *Heaven: A History*, pp. 228–29.

188 *"No marriage in heaven"*: Mark 12:25.

188 *"Into his divine arms"*: From Mechthild of Magdeburg and F. J. Tobin, *The Flowing Light of the Godhead* (New York: Paulist Press, 1998), quoted in McDannell and Lang, *Heaven: A History*, p. 101.

188 *"Should someone ask"*: S. Gaon, *The Book of Beliefs and Opinions* (New Haven, CT: Yale University Press, 1948), p. 286. With gratitude to Dov Weiss, doctoral student in the history of Judaism, University of Chicago.

188 *Begs readers to pray*: Augustine, *Confessions* (Oxford: Clarendon Press, 1992), book 9, chapter 13.

188 *Consoles a grieving widow*: Augustine, "Letter 130 (A.D. 412): To Proba, a Devoted Handmaid of God," *New Advent*, 2008, http://www.newadvent.org/fathers/1102130.htm.

189 *"Having no neighbor"*: T. Aquinas, *Summa Theologica*, vol. 2 (II–1), Q4—Art. 8, quoted in P. Hawkins, *Undiscovered Country: Imagining the World to Come* (New York: Seabury Books, 2009), p. 72.

189 *Crowded stadium*: Dante Alighieri, *Paradiso*, verse translation by Robert and Jean Hollander, introduction and notes by Robert Hollander (New York: Doubleday, 2007), cantos 31–32.

189 *"Look at Anna"*: Dante Alighieri, *Paradiso*, canto 32:133–135, p. 799.

190 *More tender and human*: I. Earls, *Renaissance Art: A Topical Dictionary* (New York: Greenwood Press, 1987), p. 173.

190 *"Irrespective of spiritual merit"*: McDannell and Lang, *Heaven: A History*, pp. 133.

190 *"Purified heaven"*: See A. E. McGrath, *Reformation Thought: An Introduction* (Oxford: Blackwell, 1999) and A. E. McGrath, *A Brief History of Heaven* (Malden, MA: Blackwell, 2003).

191 *In Puritan New England*: For a more complete discussion of the Puritans' limited recreational options, see B. C. Daniels, *Puritans at Play: Leisure and Recreation in Colonial New England* (New York: St. Martin's Press, 1995).

191 *Described the Puritan mind-set*: Thanks to David D. Hall, Bartlett Research Professor of New England Church History at Harvard Divinity School, for his guidance on early American views on heaven.

191 *"None of that Confusion"*: Sermon III in J. Mitchell, *A Discourse of the Glory to Which God Hath Called Believers by Jesus Christ Delivered in Some Sermons out of the I Pet. 5 Chap. 10 ver.: Together with an Annexed Letter* (London: Printed for Nathaniel Ponder at the Peacock in the Poultry, 1677), p. 67.

192 *"Enjoyment of persons loved"*: H. S. Stout, *The New England Soul: Preaching and*

Religious Culture in Colonial New England (New York: Oxford University Press, 1986), p. 229.

192 *The roots . . . Swedenborg*: See chapter seven ("Swedenborg and the Emergence of a Modern Heaven") in McDannell and Lang, *Heaven: A History*, pp. 181–227.

193 *Began to have visions*: Rose notes that some stories in the Swedenborgian tradition suggest that Swedenborg spoke to angels as a child.

193 *"Eat not so much"*: C. O. Sigstedt, *The Swedenborg Epic: The Life and Works of Emanuel Swedenborg* (New York: Record Press, 1971), p. 198.

194 *"Like the dwellings on earth"*: E. Swedenborg, *Heaven and Its Wonders and Hell, Drawn from Things Heard and Seen* (West Chester, PA: Swedenborg Foundation, 2000), quoted in McDannell and Lang, *Heaven: A History*, p. 192.

194 *Meeting Martin Luther*: McDannell and Lang, *Heaven: A History*, p. 188. See also W. M. White, *Emanuel Swedenborg: His Life and Writings* (London: Simpkin, Marshall, 1867), pp. 437–38.

194 *Swedenborg meets Cicero*: White, *Emanuel Swedenborg*, pp. 349–51.

195 *"I long to scatter"*: Swedenborg Foundation, "Illuminating the World of Spirit," Swedenborg.com, 2005, http://www.swedenborg.com/catalog/2005catalog.pdf.

196 *"Sublime genius"*: R. W. Emerson, "Swedenborg; or, the Mystic," in *Representative Men: Seven Lectures* (Boston and New York: Houghton, Mifflin, 1903), p. 112.

196 *"Humanizing" of heaven*: N. O. Hatch, *The Democratization of American Christianity* (New Haven, CT: Yale University Press, 1989). See also S. R. Prothero, *American Jesus: How the Son of God Became a National Icon* (New York: Farrar, Straus and Giroux, 2003).

196 *"And the mother gave"*: H. W. Longfellow, "The Reaper and the Flowers," in *The Poetical Works of Henry Wadsworth Longfellow* (Boston: Houghton Mifflin, 1986), vol. 1, pp. 22–23. Thanks to David Hall for pointing me to this poem.

197 *"I actually saw men"*: Church of Jesus Christ of Latter-day Saints, *History of the Church of Jesus Christ of Latter-day Saints*, notes by B. H. Roberts (Salt Lake City, Utah: Church of Jesus Christ of Latter-day Saints, 1948), p. 362.

197 *"The same sociality"*: Ibid., p. 323.

197 *Heaven . . . offered no comfort*: P. S. Paludan, "Religion and the American Civil War," in *Religion and the American Civil War*, ed. R. M. Miller, H. S. Stout, and C. R. Wilson (New York: Oxford University Press, 1998), p. 30.

197 *The idea of a domesticated heaven*: See McDannell and Lang, *Heaven: A History*, pp. 264–73; also note Paludan, "Religion and the American Civil War," pp. 21–42. For a related discussion of shifting views toward death and the afterlife during the Civil War, see G. Laderman, *The Sacred Remains: American Attitudes toward Death, 1799–1883,* (New Haven, CT: Yale University Press, 1999), pp. 132–33.

197 The Gates Ajar: E. S. Phelps, *The Gates Ajar* (London: Ward, Lock, and Tyler, 1872).

198 *"Mountains as we see them . . . when the wind coos"*: Phelps, *The Gates Ajar*, p. 85.

198 *"Summer holidays"*: Ibid., p. 88, quoting from C. Lamb, *Essays of Elia: To Which Are Added Letters, and Rosamund, a Tale* (Paris: Baudry's European Library, 1835), p. 32.

198 *"Blessed sunshine"*: Phelps, *The Gates Ajar*, p. 61.

198 *"See the sparkle"*: Ibid., p. 52.

198 *Sold eighty thousand copies*: McDannell and Lang, *Heaven: A History*, p. 265.

199 *"Don't believe a loving God"*: "Universalism," Unitarian Universalist Association of Congregations, April 18, 2008, http://www.uua.org/visitors/ourhistory/6904.shtml.

199 *Spiritualism started as a vogue*: I am indebted to Stephen Prothero for his guidance on American Spiritualism. See also A. Braude, *Radical Spirits: Spiritualism and Women's Rights in Nineteenth-Century America* (Bloomington: Indiana University Press, 2001); R. L. Moore, *In Search of White Crows: Spiritualism, Parapsychology, and American Culture* (New York: Oxford University Press, 1977); and B. E. Carroll, *Spiritualism in Antebellum America* (Bloomington: Indiana University Press, 1997).

199 *Margaret and Kate Fox*: A. L. Underhill, *The Missing Link in Modern Spiritualism* (New York: Thomas R. Knox, 1885).

199 *"The basest cowardice"*: A. F. Tyler, *Freedom's Ferment; Phases of American Social History to 1860* (Minneapolis: University of Minnesota Press, 1994).

200 *"Tell the Fox girls"*: "Death of Margaret Fox Kane; Youngest of the Once Celebrated Fox Sisters, Mediums," *New York Times*, March 10, 1893.

200 *Fox sisters traveled America*: E. W. Vanderhoof, *Historical Sketches of Western New York* (New York: AMS Press, 1972).

200 *One astonishing night*: "Spirit Mediums Outdone. Lively Rappings in the Academy of Music. Dr. Richmond and One of the Fox Sisters Give Exhibitions of Their Skill Before a Remarkably Responsive Crowd—Spiritualism Formally Renounced," *New York Tribune*, October 22, 1888.

200 *"Poverty and obscurity"*: "Death of Margaret Fox Kane," *New York Times*.

200 *"Willie lives"*: J. H. Baker, *Mary Todd Lincoln: A Biography* (New York: W. W. Norton, 1987), p. 220–21.

200 *Planchettes*: L. A. Long, *Rehabilitating Bodies: Health, History, and the American Civil War* (Philadelphia: University of Pennsylvania Press, 2004), p. 73.

201 *"Neurotic consequences"*: P. Tillich, *Systematic Theology* (Chicago: University of Chicago Press, 1967), quoted in McDannell and Lang, *Heaven: A History*, p. 328.

201 *"We dig a lot deeper"*: D. Van Biema, "Does Heaven Exist?" *Time*, March 24, 1997.

201 *"You were a great man"*: "Online Memorial tribute: Samuel Hicks Sr.," Chris-tianMemorials.com, 2009, http://www.christianmemorials.com/tributes/sam-uel-hicks-sr (accessed May 18, 2009).

201 *"JP will be greatly missed"*: "In Loving Memory: Johnnie Lane Parker, Sr.," Val-leyOfLife.com, 2009, http://www.valleyoflife.com/JohnnieLParkerSr/ (accessed May 18, 2009).

202 *For a 2001 poll*: "Do Pets Go to Heaven?" BeliefNet.com, 2001, http://www.beliefnet.com/Inspiration/Angels/2001/05/Do-Pets-Go-to-Heaven.aspx.

203 *"The day comes"*: "The Rainbow Bridge," Critters.com, 2008, from http://www.critters.com/rainbow-bridge.php. The poem, though anonymous, closely follows the text of "All Pets Go to Heaven," an essay published in W. Sife, *The Loss of a Pet* (New York: Howell Book House, 1998), p. 162.

203 *A 2005 Gallup poll*: L. Lyons, "Paranormal Beliefs Come (Super)Naturally to Some," Gallup.com, November 1, 2005, http://www.gallup.com/poll/19558/Paranormal-Beliefs-Come-SuperNaturally-Some.aspx.

CHAPTER NINE: IS HEAVEN BORING?

207 *Their song "Heaven"*: D. Byrne and J. Harrison, "Heaven," from *Fear Of Music*, Talking Heads album, Bleu Disque Music/Warner Chappell Music, 1979.

208 *"Heaven and Hell are both metaphors"*: D. Byrne, *The New Sins* (New York: Mc-Sweeney's, 2006).

210 *"An insight or an inspiring word"*: P. Hawkins, *Undiscovered Country: Imagining the World to Come* (New York: Seabury Books, 2009), p. 78.

210 *"Many mansions"*: John 14:2.

211 *Concern over celestial tedium*: The concept of a dynamic, changing heaven was also sketched out by seventeenth- and eighteenth-century philosophers like Gottfried Wilhelm Leibniz and Immanuel Kant; for a more complete account of the idea of progress in heaven, see C. McDannell and B. Lang, *Heaven: A History* (New York: Vintage Books, 1990), pp. 276–306.

211 *"Singing hymns"*: M. Twain, *Extract from Captain Stormfield's Visit to Heaven* (New York: Oxford University Press, 1996).

211 *"A free cruise to Alaska"*: D. Fohrman, "Why in Heaven Do We Look Forward to Heaven?" *Jewish World Review*, February 17, 2006, http://www.jewishworldreview.com/david/fohrman_sabbath5.php3.

211 *Devoid of cinematic particulars*: My gratitude to Martin Marty, Fairfax M. Cone Distinguished Service Professor Emeritus at the University of Chicago, for his exhaustive insights on the changes in American religion and religiosity in modern times. See also M. E. Marty, *Pilgrims in Their Own Land: 500 Years of Religion in America* (Boston: Little, Brown, 1984).

211 *"If I drop heaven"*: J. A. Wright, *What Makes You So Strong?: Sermons of Joy and Strength from Jeremiah A. Wright, Jr.*, ed. J. K. Ross (Valley Forge, PA: Judson Press, 1993), pp. 57–58.

213 *"It's nice and warm"*: T. Gilliam and T. Jones, *The Meaning of Life*, Celandine Films, 1983.

213 *"Heaven is totally overrated"*: J. Stein, "A Little Bit of Heaven on Earth," *Los Angeles Times*, December 21, 2007.

214 Stein wrote . . . Alcorn blogged: Stein, "A Little Bit of Heaven on Earth"; and R. Alcorn, "Joel Stein, Starbucks and Heaven," The Eternal Perspectives Blog, December 21, 2007, http://randyalcorn.blogspot.com/2007/12/joel-stein-starbucks-and-heaven.html.

215 *"Seekers"*: D. Stone, "One Nation Under God?" *Newsweek*, April 7, 2009, http://www.newsweek.com/id/192915. For more detail, see W. C. Roof, *A Generation of Seekers: The Spiritual Journeys of the Baby Boom Generation* (San Francisco: HarperSanFrancisco, 1993). Also note R. C. Fuller, *Spiritual, but Not Religious: Understanding Unchurched America* (Oxford: Oxford University Press, 2001).

215 *"A big bull ring"*: E. Hemingway and C. Baker, *Ernest Hemingway, Selected Letters, 1917–1961* (New York: Scribner, 1981), p. 165.

216 No tears in heaven: Revelation 21:4.

216 *"More talk of heaven"*: M. Ralls, "What Can We Say About the Afterlife?" *The Christian Century*, December 14, 2004.

217 *"Want to go back to college"*: L. Miller, "The Gospel of Prothero," *Newsweek*, March 12, 2007.

219 *"Motion and stillness"*: J. B. Russell, *A History of Heaven: The Singing Silence* (Princeton, NJ: Princeton University Press, 1997), p. 187.

220 *"Like this present world"*: E-mail, April 21, 2009.

220 *"Change, rescue, transformation"*: N. T. Wright, *Surprised by Hope: Rethinking Heaven, the Resurrection, and the Mission of the Church* (New York: HarperOne, 2008), p. 5.

220 Through a glass darkly: 1 Corinthians 13:12.

220 *"We will see him"*: 1 John 3:2.

221 *"God will be known"*: Book 22, chapter 30, in Augustine, *The City of God Against the Pagans* (London: Penguin, 2003), p. 1087.

221 *"For my sight"*: Dante Alighieri, *Paradiso*, verse translation by Robert and Jean Hollander, introduction and notes by Robert Hollander (New York: Doubleday, 2007), canto 33.

221 *"Perfect union of the soul"*: T. Aquinas, Q96: Art. 1, quoted in McDannell and Lang, *Heaven: A History*, p. 90.

221 *"Contemplated with wonder"*: T. Aquinas, *On the Truth of the Catholic Faith:*

Summa Contra Gentiles (Garden City, NY: Hanover House, 1995), quoted in McDannell and Lang, *Heaven: A History*, p. 90.

221 *Religious mystics:* For an account of the rise of Christian mysticism, see B. McGinn, *The Presence of God: A History of Western Christian Mysticism* (New York: Crossroad, 1992). A modern imagining of passionate mysticism can be found in R. Hansen, *Mariette in Ecstasy* (New York: E. Burlingame Books, 1991).

221 *"In the life of immortality":* G. P. E. Luttikhuizen, ed., *Paradise Interpreted: Representations of Biblical Paradise in Judaism and Christianity* (Leiden: E. J. Brill, 1999), p. 177.

222 *"Took her into his arms":* From "The Herald of Divine Love," quoted in McDannell and Lang, *Heaven: A History*, p. 103. See their discussion of Gertrude, pp. 102–6. See also McDannell and Lang on the beatific vision, pp. 88–94.

222 *"Like the bow in a cloud":* Ezekiel 1:28.

222 *"Gaze upon the King":* A. F. Segal, *Life After Death: A History of the Afterlife in the Religions of the West* (New York: Doubleday, 2004), p. 508.

222 *The following account:* "The Measure of the Divine Body," Work of the Chariot, 2009, http://www.workofthechariot.com. See also W. Bacher and L. Blau, "Shi'ur Komah," in *The Jewish Encyclopedia*, 2002, www.jewishencyclopedia.com/view.jsp?artid=646&letter=S&search=Sefer%20Raziel%20ha-Gadol.

223 *"In joy . . . from the fire":* *The Zohar*, trans. H. Sperling and M. Simon (London: Soncino Press, 1934), vol. 5, p. 26, quoted in S. P. Raphael, *Jewish Views of the Afterlife* (Northvale, NJ: Jason Aronson, 1994).

223 *"What the whirling is":* S. Can, *Fundamentals of Rumi's Thought: A Mevlevi Sufi Perspective* (Istanbul: The Light, 2004), pp. 205–206. See also J. a.-D. Rûmî and C. Barks, *The Essential Rumi* (San Francisco: Harper, 1995).

223 *"His voice was deep and rich":* C. S. Lewis, *The Lion, the Witch and the Wardrobe; a Story for Children* (New York: Macmillan, 1950), pp. 127–28.

224 *"Mauled and licked clean":* M. Goldberg, *Bee Season: A Novel* (New York: Random House, 2000), p. 270.

224 *Miriam . . . has a tambourine:* Exodus 15:20.

224 *Sing praises to the Lord:* Psalms 98:6.

225 *Something angels did in heaven:* M. Weinfeld, *Normative and Sectarian Judaism in the Second Temple Period* (London: T. & T. Clark, 2005), p. 48.

225 *Full of singing:* "Holy, Holy, Holy," Revelation 4:8 (and see also Isaiah 6:3); "ten thousand times ten thousand" angels: Revelation 5:8–13.

225 *"Above, the hosts of angels":* "Homily on Isaiah," quoted in M. Barker, *The Great High Priest: The Temple Roots of Christian Liturgy* (Edinburgh: T. & T. Clark, 2003), p. 143.

226 *"It exists before they were there":* J. Ratzinger, " 'In the Presence of the Angels

I Will Sing Your Praise': The Regensburg Tradition and the Reform of the Liturgy," *Adoremus Bulletin*, October–December 1996, http://www.adoremus. org/10–12–96-Ratzi.html.

226   *Fearsome organ music*: R. Moody, "On Celestial Music," in R. Ferlo, *Heaven* (New York: Seabury Books, 2007), pp. 46–58.

226   *"Listens with special pleasure"*: K. Barth, "Wolfgang Amadeus Mozart," in *Religion and Culture: Essays in Honor of Paul Tillich*, ed. W. Leibrecht (New York: Harper, 1959), pp. 61–79. I have here followed the translation given in C. J. Green, *Karl Barth: Theologian of Freedom* (Minneapolis: Fortress Press, 1991; first published 1989 by Collins).

226   *Choral singing*: S. J. White, *Foundations of Christian Worship* (Louisville, KY: Westminster John Knox Press, 2006), p. 43.

226   *Harmony singing*: B. B. Patterson, "Appalachian Religious Music," in *Encyclopedia of Religion in the South*, ed. S. S. Hill and C. H. Lippy (Macon, GA: Mercer University Press, 2005), pp. 69–73.

229   *Fastest-growing brand*: The Pew Forum on Religion and Public Life, "Pentecostalism," PewForum.org, 2009, http://pewforum.org/docs/?DocID=140. See also "World Christian Database," Gordon-Conwell Theological Seminary, 2005, http://www.worldchristiandatabase.org.

230   *And time itself*: This, of course, was the point made by the Church Fathers who sought to collapse time: for them, eternity could not be dull, because it lacked duration in any meaningful sense. See Augustine, *Confessions*, book 11, chapter 13.

231   *By the nineteenth century*: Note, however, that some revisionist historians argue that the concept of progress was well established in classical antiquity. See, for instance, L. Edelstein, *The Idea of Progress in Classical Antiquity* (Baltimore: Johns Hopkins University Press, 1967).

231   *Occasional days off*: See section 17 in E. Swedenborg, *The Delights of Wisdom Respecting Conjugal Love: After Which Follow the Pleasures of Insanity Respecting Scortatory Love* (London: Printed for the Society, 1790).

231   *"So much work"*: McDannell and Lang, *Heaven: A History*, p. 315. See also pp. 201–11 and pp. 276–306 for a discussion of the evolution of the idea of progress in heaven.

231   *"Higher and still higher"*: J. Dodsworth, *The Better Land; or, the Christian Emigrant's Guide to Heaven* (London: Ilkeston, 1853), p. 290, quoted in McDannell and Lang, *Heaven: A History*, p. 284.

232   *"Their progress will be unlimited"*: Quoted in G. Ahmad, *The Essence of Islam* (Tilford, UK: Islam International Publications, 2004), p. 424.

232   *"To the Academy"*: H. Freedman and G. Scholem, "Academy on High," in *Ency-*

clopaedia Judaica, ed. M. Berenbaum and F. Skolnik (Detroit: Macmillan Reference USA, 2007), vol. 1, pp. 353–54.

232 *"There is no eating"*: Babylonian Talmud, Tractate Berakoth 17a, quoted in B. HaLevi, "Life After Life: Jewish Sources of Dying, Death and Beyond," RabbiB.com, 2009, http://www.rabbib.com/images/Life_After_Death_Sources.pdf.

232 *Dylan decided to study*: "Bob Dylan's Life with the Lubavitchers," *New York Magazine*, June 6, 1983.

233 *Reincarnation . . . in Jewish thought*: Raphael, *Jewish Views of the Afterlife*. I am grateful to Simcha Raphael for his analysis and insight into contemporary mystical Judaism.

234 *"All souls"*: The *Zohar*, vol. 3, p. 302, quoted in Raphael, *Jewish Views of the Afterlife*, p. 316.

236 *"U2 . . . like Bono"*: J. Pareles, "Pop and Jazz Guide," *New York Times*, June 29, 2001, and J. Pareles, "A Rock Star's Struggle Where Militant Islam Rules," *New York Times*, July 17, 2003.

236 *Who dreamed the American dream*: See S. Ahmad, *Rock and Roll Jihad: A Muslim Rock Star's Revolution for Peace* (New York: Free Press, 2010).

238 *Tawhid*: See also H. A. Hameed, "Tauhid and Adl: A Discussion," in *Encyclopaedic Survey of Islamic Culture*, ed. M. Taher (New Delhi: Anmol Publications, 1997), pp. 77–101.

EPILOGUE

242 *"Gathered to his people"*: Genesis 25:8.

243 *Whose works I generally admire*: See, for example, S. Harris, *The End of Faith: Religion, Terror, and the Future of Reason* (New York: W. W. Norton, 2004), and C. Hitchens, *God Is Not Great: How Religion Poisons Everything* (New York: Grand Central, 2007).

243 *Just a third*: "While Most U.S. Adults Believe in God, Only 58 Percent Are 'Absolutely Certain,' " Harris Interactive, October 31, 2006, http://www.harrisinteractive.com/harris_poll/index.asp?PID=707.

244 *"I go out at night"*: V. van Gogh, "Letter to Theo van Gogh, 28 September 1888," The Letters of Vincent van Gogh, No. 543, trans. R. Harrison, http://www.webexhibits.org/vangogh/letter/18/543.htm.

244 *Agnes Long*: L. Miller, "Life in Solitary," *Newsweek*, June 20, 2005. Agnes Long has since left Madeline Island and can't be located.

246 *Heaven is marriage*: See Revelation 21:2.

247 *"Loves everybody"*: J. Meacham, "Pilgrim's Progress," *Newsweek*, August 14, 2006.

248 *What we cannot reach*: E. Dickinson, *The Poems of Emily Dickinson: Reading Edition*, ed. R. W. Franklin (Cambridge, MA: Belknap Press of Harvard University Press, 1999).

BIBLIOGRAPHY

Abdesselem, M. *Le thème de la mort dans la poésie arabe des origines à la fin du IIIe/IXe siècle*. Tunis, Tunisia: Publications de l'Université de Tunis, 1977.

Abou El Fadl, K. "What Became of Tolerance in Islam?" *Los Angeles Times*, September 14, 2001.

Ahmad, G. *The Essence of Islam*. Tilford, UK: Islam International Publications, 2004.

Ahmad, S. *Rock and Roll Jihad: A Muslim Rock Star's Revolution for Peace*. New York: Free Press, 2010.

Ahmadinejad, Mahmoud. Transcript of U.N. speech. NPR.org. Retrieved April 20, 2009, from www.npr.org/templates/story/story.php?storyId=6107339.

Aichele, G. *Jesus Framed*. London: Routledge, 1996.

Albom, M. *The Five People You Meet in Heaven*. New York: Hyperion, 2003.

Alcorn, R. C. *Edge of Eternity*. Waterville, ME: Five Star, 2003.

———. "Joel Stein, Starbucks and Heaven." The Eternal Perspectives Blog, December 21, 2007. Retrieved May 18, 2009, from http://randyalcorn.blogspot.com/2007/12/joel-stein-starbucks-and-heaven.html.

Ali, A. Yusuf. *The Qur'an: Translation*. New York: Tahrike Tarsile Qur'an, 2007.

Althaus, P. *The Theology of Martin Luther*. Philadelphia: Fortress Press, 1966.

Amanat, A., and M. T. Bernhardsson. *Imagining the End: Visions of Apocalypse from the Ancient Middle East to Modern America*. London: I. B. Tauris, 2002.

Andrews, H. T. *The Apocryphal Books of the Old and New Testament*. London: T. C. & E. C. Jack, 1908.

Aquinas, T. *On the Truth of the Catholic Faith: Summa Contra Gentiles*. Garden City, NY: Hanover House, 1995.

———. *Summa Theologica*. Vol. 5, pt. 3, 2nd sect. and suppl. New York: Cosimo, 2007.

Arasse, D., and A. Kiefer. *Anselm Kiefer*. New York: Harry N. Abrams, 2001.

Arberry, A. J. *The Koran Interpreted*. New York: Touchstone, 1996.

Armstrong, K. A *History of God: From Abraham to the Present, the 4,000-Year Quest for God*. London: Heinemann, 1993.

———. *Muhammad: A Biography of the Prophet*. San Francisco: HarperSanFrancisco, 1993.

———. *Islam: A Short History*. New York: Modern Library, 2002.

Arnold-Baker, C. *The Companion to British History*. London: Routledge, 2001.

Arrington, L. J., and D. Bitton. *The Mormon Experience: A History of the Latter-Day Saints*, 2nd ed. Urbana: University of Illinois Press, 1992.

Ashmore, R. "McVeigh Will Meet Unimaginable Mercy." *National Catholic Reporter*, May 18, 2001.

Associated Press. "Diamonds Can't Tempt Honest LA Cabbie." *MSNBC*, November 18, 2005. Retrieved April 22, 2009, from http://www.msnbc.msn.com/id/10096893/.

Associated Press. "Obama Reaches Out to Christian Leaders." *MSNBC*, June 11, 2008. Retrieved April 30, 2009, from http://www.msnbc.msn.com/id/25092483/.

Athanasius. "Life of Anthony." In *A Select Library of Nicene and Post-Nicene Fathers of the Christian Church*, edited by P. Schaff and H. Wace, 4:188–221. New York: Charles Scribner's Sons, 1903.

Augustine. *City of God*. Trans. H. Bettenson. London: Penguin, 2003.

———. "The City of God Against the Pagans." In *A Select Library of the Nicene and Post-Nicene Fathers of the Christian Church*, edited by P. Schaff, 2:1–511. New York: Charles Scribner's Sons, 1899.

———. *The City of God Against the Pagans*. Cambridge: Cambridge University Press, 1998.

———. *Confessions*. Oxford: Clarendon Press, 1992.

———. "Letter 130 (A.D. 412): To Proba, a Devoted Handmaid of God." *New Advent*. Retrieved May 18, 2009, from http://www.newadvent.org/fathers/1102130.htm.

———. "The Magnitude of the Soul." In *The Fathers of the Church*. Washington, D.C.: Catholic University of America Press, 1997.

———. "On Merit and the Forgiveness of Sins, and the Baptism of Infants." *New Advent*. Retrieved April 30, 2009, from http://www.newadvent.org/fathers/15011.htm.

Bacher, W., and L. Blau. "Shi'ur Komah." In *The Jewish Encyclopedia*, 2002. Retrieved May 18, 2009, from http://www.jewishencyclopedia.com/view.jsp?artid=646&letter=S&search=Sefer%20Raziel%20ha-Gadol.

Bair, D. *Samuel Beckett: A Biography*. New York: Harcourt Brace Jovanovich, 1978.

Baker, J. H. *Mary Todd Lincoln: A Biography*. New York: W. W. Norton, 1987.

Bakhtiar, L. *Encyclopedia of Islamic law: A Compendium of the Views of the Major Schools*. Chicago: ABC International Group, 1996.

———. *The Sublime Quran*. Chicago: Kazi Publications, 2007.

Banks, A. M., and D. Burke. "Fuller Picture Emerges of Obama's Evangelical Meeting." Religion News Service, June 18, 2008. Retrieved April 30, 2009, from http://pewforum.org/news/display.php?NewsID=15867.

Barker, M. *The Great High Priest: The Temple Roots of Christian Liturgy*. Edinburgh: T. and T. Clark, 2003.

Barnhart, R. K., and S. Steinmetz. *Chambers Dictionary of Etymology: The Origins and Development of Over 25,000 English Words*. Edinburgh: Chambers, 2006.

Barrett, P. *American Islam: The Struggle for the Soul of a Religion*. New York: Farrar, Straus and Giroux, 2007.

Barth, K. "Wolfgang Amadeus Mozart." In *Religion and Culture: Essays in Honor of Paul Tillich*, edited by W. Leibrecht, 61–79. New York: Harper, 1959.

Baumgartner, F. J. *Longing for the End: A History of Millennialism in Western Civilization*. New York: Palgrave, 2001.

Belanti, J., M. Perera, and K. Jagadheesan. "Phenomenology of Near-Death Experiences: A Cross-Cultural Perspective." *Transcultural Psychiatry* 45, no. 1 (2008): 121–33.

BeliefNet.com. "Do Pets Go to Heaven?" Retrieved May 17, 2009, from http://www.beliefnet.com/Inspiration/Angels/2001/05/Do-Pets-Go-to-Heaven.aspx.

Benedict. *Rule of Saint Benedict in English*. Collegeville, MN: Liturgical Press, 1981.

Berger, D. *The Rebbe, the Messiah, and the Scandal of Orthodox Indifference*. London: Littman Library of Jewish Civilization, 2001.

Bernstein, L. *West Side Story*. Book by Arthur Laurents. Music by Leonard Bernstein. Lyrics by Stephen Sondheim and Leonard Bernstein. London: Heinemann, 1959.

Bickerman, E. J., and Jewish Theological Seminary of America. *The Jews in the Greek Age*. Cambridge, MA: Harvard University Press, 1988.

Bivar, A. D. H. "Achaemenid Coins, Weights and Measures." In *The Cambridge History of Iran: The Median and Achamenian Periods*, edited by I. Gershevich, 610–35. Cambridge: Cambridge University Press, 1985.

Blair, S., and J. Bloom, eds. *Images of Paradise in Islamic Art*. Hanover, NH: Hood Museum of Art, Dartmouth College, 1991.

Bleich, J. D., ed. *With Perfect Faith: The Foundations of Jewish Belief*. New York: Ktav, 1982.

Bloch-Smith, E. *Judahite Burial Practices and Beliefs About the Dead*. Sheffield, UK: JSOT Press, 1992.

Bloom, H. *The Western Canon: The Books and School of the Ages*. New York: Riverhead Trade, 1995.

Blue Letter Bible. "Lexicon Results." Retrieved May 1, 2009, from http://www.blueletterbible.org/lang/lexicon/lexicon.cfm?Strongs=H6663&t=KJV.

"Bob Dylan's Life with the Lubavitchers." *New York Magazine*, June 6, 1983.

Boccaccio, G., and L. B. Aretino. *The Earliest Lives of Dante.* Translated by James Robinson Smith. New York: Henry Holt, 1901.

Botticelli, S. *Sandro Botticelli: The Drawings for Dante's Divine Comedy.* London: Royal Academy of Arts, 2000.

Boyer, P. S. *When Time Shall Be No More: Prophecy Belief in Modern American Culture.* Cambridge, MA: Belknap Press of Harvard University Press, 1992.

Braude, A. *Radical Spirits: Spiritualism and Women's Rights in Nineteenth-Century America.* Bloomington: Indiana University Press, 2001.

Brecht, M. *Martin Luther: His Road to Reformation, 1483–1521.* Minneapolis: Fortress Press, 1993.

Bremmer, J. N. *The Early Greek Concept of the Soul.* Princeton, NJ: Princeton University Press, 1983.

———. *The Rise and Fall of the Afterlife: The 1995 Read-Tuckwell Lectures at the University of Bristol.* London: Routledge, 2002.

Brodie, T. L. *Genesis as Dialogue: A Literary, Historical, and Theological Commentary.* Oxford: Oxford University Press, 2001.

Bromiley, G. W. *International Standard Bible Encyclopedia.* Grand Rapids, MI: William B. Eerdmans, 1979.

Brown, M. "About the Author." LatterDayLogic.com. Retrieved April 15, 2009, from http://www.latterdaylogic.com/about/.

Brown, P. R. L. *Augustine of Hippo: A Biography.* Berkeley: University of California Press, 2000.

———. *The Body and Society: Men, Women, and Sexual Renunciation in Early Christianity.* New York: Columbia University Press, 1988.

———. *The Cult of the Saints: Its Rise and Function in Latin Christianity.* Chicago: University of Chicago Press, 1981.

Bultmann, R. "The Message of Jesus and the Problem of Mythology." In *The Historical Jesus in Recent Research,* edited by J. D. G. Dunn and S. McKnight, 531–42. Winona Lake, IN: Eisenbrauns, 2005.

Burke, E. *Pleasure and Pain: Reminiscences of Georgia in the 1840s.* Savannah: Beehive Press, 1978.

Burkert, W. " 'Or Also a Godly Singer': Akkadian and Early Greek Literature." In *Gilgamesh: A Reader,* edited by J. Maier, 178–91. Wauconda, IL: Bolchazy-Carducci Publishers, 1997.

Burton-Christie, D. *The Word in the Desert.* New York: Oxford University Press, 1993.

Bushman, C. L., and R. L. Bushman. *Building the Kingdom: A History of Mormons in America.* Oxford: Oxford University Press, 2001.

Byassee, J. "Africentric Church: A Visit to Chicago's Trinity UCC." *The Christian Century,* May 29, 2007.

Bynum, C. W. *The Resurrection of the Body in Western Christianity, 200–1336.* New York: Columbia University Press, 1995.

Byrne, D. *The New Sins.* New York: McSweeney's, 2006.

Byrne, D., and J. Harrison. "Heaven." In *Fear of Music*, album. Bleu Disque Music/ Warner Chappell Music, 1979.

Byrne, R. *The Secret.* New York: Atria Books, 2006

Camping, H. "We Are Almost There!" Family Stations, Inc. Retrieved April 25, 2009, from http://www.familyradio.com/graphical/literature/waat/waat.pdf.

Can, Ş. *Fundamentals of Rumi's Thought: A Mevlevi Sufi Perspective.* Istanbul: The Light, 2004.

Capps, D. *A Time to Laugh: The Religion of Humor.* New York: Continuum, 2005.

Carnes, T., and A. Karpathakis, eds. *New York Glory: Religions in the City.* New York: New York University Press, 2001.

Carroll, B. E. *Spiritualism in Antebellum America.* Bloomington: Indiana University Press, 1997.

Cassara, E. *Universalism in America: A Documentary History.* Boston: Beacon Press, 1971.

Central Conference of American Rabbis. "Pittsburgh Platform: Declaration of Principles." In *Religion and American Cultures*, edited by G. Laderman and L. D. León, 779. Santa Barbara, CA: ABC-CLIO, 2003.

Chadwick, O. *A History of Christianity.* London: Weidenfeld & Nicolson, 1995.

Charlesworth, J. H., and C. D. Elledge. *Resurrection: The Origin and Future of a Biblical Doctrine.* New York: T. & T. Clark, 2006.

Chaucer, G. *The Canterbury Tales.* Mineola, NY: Dover, 2004.

ChristianMemorials.com. "Online Memorial tribute: Samuel Hicks Sr." *ChristianMemorials.com.* Retrieved May 18, 2009, from http://www.christianmemorials .com/tributes/samuel-hicks-sr.

Chryssavgis, J. *In the Heart of the Desert: The Spirituality of the Desert Fathers and Mothers.* Bloomington, IN: World Wisdom, 2003.

Church of Jesus Christ of Latter-day Saints. *History of the Church of Jesus Christ of Latter-day Saints.* Notes by B. H. Roberts. Salt Lake City, Utah: Church of Jesus Christ of Latter-day Saints, 1948.

———. "The Official Scriptures of the Church of Jesus Christ of Latter-day Saints." LDS.org. Retrieved May 12, 2009, from http://scriptures.lds.org/en/contents.

Clark, P. *Zoroastrianism: An Introduction to an Ancient Faith.* Brighton: Sussex Academic Press, n.d.

Clement of Alexandria. "Exhortation to the Heathen." In B. M., Metzger, *The Canon of the New Testament: Its Origin, Development, and Significance.* Oxford: Clarendon Press, 1997.

Cochrane, A. C., ed. *Reformed Confessions of the 16th Century.* Louisville, KY: Westminster John Knox Press, 2003.

Cohen, N., and D. Cohen. *Long Steel Rail: The Railroad in American Folksong.* Urbana: University of Illinois Press, 2000.

Colby, F. S. *Narrating Muhammad's Night Journey: Tracing the Development of the Ibn 'Abbâs Ascension Discourse.* Albany: State University of New York Press, 2008.

Coleridge, S. T. *Biographia Literaria.* London: G. Bell and Sons, 1898.

Collins, A. Y. *Crisis and Catharsis: The Power of the Apocalypse.* Philadelphia: Westminster Press, 1984.

Collins, J. J. *Daniel: With an Introduction to Apocalyptic Literature.* Grand Rapids, MI: William B. Eerdmans, 1984.

Collins, O., ed. *Speeches That Changed the World.* Louisville, KY: Westminster John Knox Press, 1999.

Collinson, P. *The Reformation.* London: Weidenfeld & Nicolson, 2003.

Cornell, V. J. *Voices of Islam.* Westport, CT: Praeger, 2007.

Cottret, B. *Calvin: A Biography.* Grand Rapids, MI: William B. Eerdmans, 1995.

Cox-Chapman, M. *The Case for Heaven: Near-Death Experiences as Evidence of the Afterlife.* New York: G. P. Putnam's Sons, 1995.

Cremation Association of North America. "Final 2005 Statistics and Projections to the Year 2025; 2006 Preliminary Data." 89th Annual Convention of the Cremation Association of North America. San Francisco, 2007.

Critters.com. "The Rainbow Bridge." *Critters.com.* Retrieved May 28, 2009, from http://www.critters.com/rainbow-bridge.php.

Crone, P. *From Arabian Tribes to Islamic Empire: Army, State and Society in the Near East c. 600–850.* Aldershot, UK: Ashgate, 2008.

———. *Meccan Trade and the Rise of Islam.* Princeton, NJ: Princeton University Press, 1987.

Crossan, J. D. *Jesus: A Revolutionary Biography.* San Francisco: HarperSanFrancisco, 1994.

———. *The Historical Jesus: The Life of a Mediterranean Jewish Peasant.* San Francisco: HarperSanFrancisco, 1991.

D'Aquili, E., and A. B. Newberg. *The Mystical Mind: Probing the Biology of Religious Experience.* Minneapolis: Augsburg Fortress, 1999.

Daniell, C. *Death and Burial in Medieval England, 1066–1550.* London: Routledge, 1997.

Daniels, B. C. *Puritans at Play: Leisure and Recreation in Colonial New England.* New York: St. Martin's Press, 1995.

Dante Alighieri. *Paradiso.* Verse translation by Robert and Jean Hollander. Introduction and notes by Robert Hollander. New York: Doubleday, 2007.

Davidson, C. *The Iconography of Heaven.* Kalamazoo: Medieval Institute Publications, Western Michigan University, 1994.

Davidson, H. A. *Moses Maimonides: The Man and His Works.* New York: Oxford University Press, 2005.

"Death of Margaret Fox Kane; Youngest of the Once Celebrated Fox Sisters, Mediums." *New York Times*, March 10, 1893.

Deshman, R. "Another Look at the Disappearing Christ: Corporeal and Spiritual Vision in Early Medieval Images." *The Art Bulletin* 79, no. 3 (1997): 518–46.

Desrosiers, G. *An Introduction to Revelation*. London: Continuum International Publishing Group, 2000.

Diamant, A. *Saying Kaddish: How to Comfort the Dying, Bury the Dead, and Mourn as a Jew*. New York: Schocken, 1998.

Dickey, C. *Summer of Deliverance: A Memoir of Father and Son*. New York: Touchstone, 1998.

Dickinson, E. *The Poems of Emily Dickinson: Reading Edition*. Edited by R. W. Franklin. Cambridge, MA: Belknap Press of Harvard University Press, 1999.

Dodsworth, J. *The Better Land; or, the Christian Emigrant's Guide to Heaven*. London: Ilkeston, 1853.

Donner, F. M. "The Background to Islam." In *The Cambridge Companion to the Age of Justinian*, ed. M. I. Maas, 510–34. New York: Cambridge University Press, 2005.

Doughty, C. M. *Travels in Arabia Deserta*. Cambridge: Cambridge University Press, 1888.

Drews, R. "Judaism, Christianity, and Islam to the Beginnings of Modern Civilization." Vanderbilt University, April 20, 2009. Retrieved April 20, 2009, from http://site mason.vanderbilt.edu/classics/drews/COURSEBOOK.

Duffy, E. *The Stripping of the Altars: Traditional Religion in England, c. 1400–c. 1580*. New Haven, CT: Yale University Press, 2005.

Duke, J. T. "Marriage: Eternal Marriage." In *Encyclopedia of Mormonism*, ed. D. H. Ludlow. New York: Macmillan, 1992.

Earls, I. *Renaissance Art: A Topical Dictionary*. New York: Greenwood Press, 1987.

Edelstein, L. *The Idea of Progress in Classical Antiquity*. Baltimore: Johns Hopkins University Press, 1967.

El Fadl, K. Abou. *The Great Theft: Wrestling Islam from the Extremists*. New York: HarperCollins, 2005.

Elior, R. *Memory and Oblivion: The Secret of the Dead Sea Scrolls*. Jerusalem: Van Leer Institute and ha-Kibutz ha-Meuchad, 2009.

Elliott, J. K. *The Apocryphal New Testament: A Collection of Apocryphal Christian Literature in an English Translation*. Oxford: Oxford University Press, 1993.

Emerson, J. S., and H. Feiss. *Imagining Heaven in the Middle Ages: A Book of Essays*. New York: Garland, 2000.

Emerson, R. W. "Swedenborg; or, the Mystic." In *Representative Men: Seven Lectures*. Boston and New York: Houghton, Mifflin, 1903.

———. "Concord, October 30, 1840." In *The Correspondence of Thomas Carlyle and Ralph Waldo Emerson, 1834–1872*. Vol. 1. Cambridge, MA: Riverside Press, 1896.

The Encyclopedia of Apocalypticism. Edited by B. McGinn, J. J. Collins, and S. J. Stein. New York: Continuum, 1998.

European Commission. Special Eurobarometer 225: Social Values, Science and Technology. Brussels, Belgium, 2005.

Evagrius. "On the Eight Thoughts." In Evagrius of Pontus: The Greek Ascetic Corpus, edited by R. E. Sinkewicz, 66–90. Oxford: Oxford University Press, 2006.

Falwell, J. "Tinky Winky Comes Out of the Closet." NLJ Online. Retrieved April 14, 2009, from http://web.archive.org/web/19990423025753/; www.liberty.edu/chancellor/nlj/feb99/politics2.htm.

Feinberg, J. S. No One Like Him: The Doctrine of God. Wheaton, IL: Crossway Books, 2006.

Ferlo, R. Heaven. New York: Seabury Books, 2007.

Ferrante, J. M. "Beatrice." in The Dante Encylopedia, ed. R. Lansing. New York: Garland Publishing, 2000.

Filiu, J.-P. L'apocalypse dans l'Islam. Paris: Fayard, 2008.

Fitzgerald, A., and J. C. Cavadini. Augustine Through the Ages: An Encyclopedia. Grand Rapids, MI: William B. Eerdmans, 1999.

Flannery, T. "Billy Graham's Apostasy." WorldNetDaily, August 10, 2006. Retrieved April 30, 2009, from http://www.worldnetdaily.com/news/article.asp?ARTICLE_ID=51461.

Fohrman, D. "Why in Heaven Do We Look Forward to Heaven?" Jewish World Review, February 17, 2006. Retrieved May 18, 2009, from http://www.jewishworldreview.com/david/fohrman_sabbath5.php3.

Ford, H. "St. Benedict of Nursia." In The Catholic Encyclopedia, edited by Charles G. Herbermann, 2. New York: Robert Appleton Company, 1907.

Frank, G. The Memory of the Eyes: Pilgrims to Living Saints in Christian Late Antiquity. Berkeley: University of California Press, 2000.

Fredriksen, P. Augustine and the Jews: A Christian Defense of Jews and Judaism. New York: Doubleday, 2008.

———. From Jesus to Christ: The Origins of the New Testament Images of Jesus. New Haven, CT: Yale University Press, 1988.

Freedman, D. N., and M. J. McClymond. The Rivers of Paradise: Moses, Buddha, Confucius, Jesus, and Muhammad as Religious Founders. Grand Rapids, MI: William B. Eerdmans, 2001.

Freedman, H., and G. Scholem. "Academy on High." In Encyclopaedia Judaica, ed. M. Berenbaum and F. Skolnik, 1:353–54. Detroit: Macmillan Reference USA, 2007.

Friedenberg, R. V. Notable Speeches in Contemporary Presidential Campaigns. Westport, CT: Praeger, 2002.

Friedman, M. Doesn't Anyone Blush Anymore?: Reclaiming Intimacy, Modesty and Sexuality. San Francisco: HarperCollins, 1990.

Friedman, R. E. *Who Wrote the Bible?* New York: Summit Books, 1987.

Friesen, S. J. *Imperial Cults and the Apocalypse of John: Reading Revelation in the Ruins.* Oxford: Oxford University Press, 2001.

Fugal, E. W. "Salvation of the Dead." In *Encyclopedia of Mormonism*, ed. D. H. Ludlow. New York: Macmillan, 1992.

Fuller, R. C. *Spiritual, but Not Religious: Understanding Unchurched America.* Oxford: Oxford University Press, 2001.

Furnish, T. R. *Holiest Wars: Islamic Mahdis, Their Jihads, and Osama bin Laden.* Westport, CT: Praeger, 2005.

Gafni, I. "Antiochus." In *Encyclopaedia Judaica*, ed. M. Berenbaum and F. Skolnik, 2:202–204. Detroit: Macmillan Reference USA, 2007.

Gallup, Inc. "Gallup/Nathan Cummings Foundation and Fetzer Institute Poll." Gallup, Inc. Retrieved April 13, 2009, from www.gallup.com/poll/1690/Religion .aspx.

Gallup, G., and W. Proctor. *Adventures in Immortality.* London: Corgi Books, 1984.

Gaon, S. *The Book of Beliefs and Opinions.* New Haven, CT: Yale University Press, 1948.

Gardiner, E. "Bibliography on St Patrick's Purgatory." *Hell-On-Line.* Retrieved May 13, 2009, from www.hell-on-line.org/BibPatrick.html.

———. *Medieval Visions of Heaven and Hell: A Sourcebook.* New York: Garland, 1993.

———. *Visions of Heaven and Hell Before Dante.* New York: Italica Press, 1988.

Garthwaite, G. R. *The Persians.* Malden, MA: Blackwell, 2004.

Gass, W. H. *Reading Rilke: Reflections on the Problems of Translation.* New York: Alfred A. Knopf, 1999.

Geary, P. J. "Penance (The West)." In *Encyclopedia of the Middle Ages* (e-reference ed.), ed. A. Vauchez, trans. A. Walford. Chicago: Fitzroy Dearborn, 2000. Retrieved May 25, 2009, from http://www.oxford-middleages.com/entry?entry=t179 .e2155-sl.

Gerrish, B. A. "The Place of Calvin in Christian Theology." In *The Cambridge Companion to John Calvin*, ed. D. K. McKim, 289–304. Cambridge: Cambridge University Press, 2004.

Gibbs, N. R., and M. Duffy. *The Preacher and the Presidents: Billy Graham in the White House.* New York: Center Street, 2007.

Gillman, N. *The Death of Death: Resurrection and Immortality in Jewish Thought.* Woodstock, VT: Jewish Lights, 1997.

Gingerich, O. *God's Universe.* Cambridge, MA: The Belknap Press of Harvard University Press, 2006.

Glassé, C. "Arabs." In *The New Encyclopedia of Islam*, 58–60. Walnut Creek, CA: AltaMira Press, 2002.

Goldberg, M. *Bee Season: A Novel.* New York: Random House, 2000.

Goller, H. "Israeli Inquiry Moves to Hebron Massacre Site." Reuters News, March 10, 1994.

Goodman, M. *The Ruling Class of Judaea: The Origins of the Jewish Revolt Against Rome A.D. 66–70.* Cambridge: Cambridge University Press, 1987.

Goodyear, D. "Quiet Depravity." *New Yorker*, October 24, 2005.

Gordon-Conwell Theological Seminary. *World Christian Database.* Retrieved May 26, 2009, from http://www.worldchristiandatabase.org.

Grant, E. *Physical Science in the Middle Ages.* Cambridge: Cambridge University Press, 1977.

———. *Planets, Stars, and Orbs: The Medieval Cosmos, 1200–1687.* Cambridge: Cambridge University Press, 1994.

Green, A. R. W. *The Storm-God in the Ancient Near East.* Winona Lake, IN: Eisenbrauns, 2003.

Green, C. J. *Karl Barth: Theologian of Freedom.* Minneapolis: Fortress Press, 1991. First published 1989 by Collins.

Gregory Nazianzen. "Select Orations of Saint Gregory Nazianzen." In *A Select Library of Nicene and Post-Nicene Fathers of the Christian Church* (2nd series), ed. P. Schaff and H. Wace, 7:203–436. Grand Rapids, MI: T. & T. Clark.

Gregory. *Morals on the Book of Job.* Oxford: F. and J. Rivington, 1850.

Greyson, B. "Dissociation in People Who Have Near-Death Experiences: Out of Their Bodies or Out of Their Minds?" *The Lancet* 355 (2000): 460–63.

Greyson, B. "The Incidence of Near-Death Experiences." *Medicine and Psychiatry* 1 (1998): 92–99.

Griffin, E. *The Cloud of Unknowing.* New York: HarperOne, 2004.

Griffler, K. P. *Front Line of Freedom: African Americans and the Forging of the Underground Railroad in the Ohio Valley.* Lexington: University Press of Kentucky, 2004.

Gruber, M. *Journey Back to Eden: My Life and Times Among the Desert Fathers.* Maryknoll, NY: Orbis Books, 2004.

Guttman, N. "McCain Battles for Credibility with Jews." *The Jewish Chronicle* (Washington, DC), March 14, 2008.

Hachlili, R. *Ancient Jewish Art and Archaeology in the Land of Israel.* Leiden: E. J. Brill, 1988.

Haddad, Y. Y., and J. I. Smith. *Muslim Communities in North America.* Albany: State University of New York Press, 1994.

HaLevi, B. "Life After Life: Jewish Sources of Dying, Death and Beyond." RabbiB.com. Retrieved May 18, 2009, from http://www.rabbib.com/images/Life_After_Death_Sources.pdf.

Halevi, L. *Muhammad's Grave: Death Rites and the Making of Islamic Society.* New York: Columbia University Press, 2007.

Hallote, R. S. *Death, Burial, and Afterlife in the Biblical World: How the Israelites and Their Neighbors Treated the Dead*. Chicago: Ivan R. Dee, 2001.

Hameed, H. A. "Tauhid and Adl: A Discussion." In *Encyclopaedic Survey of Islamic Culture*, edited by M. Taher, 77–101. New Delhi: Anmol Publications, 1997.

Hammer, R. *Entering the High Holy Days: A Guide to Origins, Themes, and Prayers*. Philadelphia: Jewish Publication Society of America, 2005.

Hansen, R. *Mariette in Ecstasy*. New York: E. Burlingame Books, 1991.

Harris, B. "Reform Siddur Revives Resurrection Prayer." Jewish Telegraph Agency, November 20, 2007.

Harris, S. *The End of Faith: Religion, Terror, and the Future of Reason*. New York: W. W. Norton, 2004.

Harris, S., and A. Sullivan. "Is Religion 'Built Upon Lies'?" *BeliefNet*, January 23, 2007. Retrieved April 28, 2009, from http://www.beliefnet.com/Faiths/Secular-Philosophies/Is-Religion-Built-Upon-Lies.aspx.

Harris Interactive. "More Americans Believe in the Devil, Hell and Angels Than in Darwin's Theory of Evolution."http://www.harrisinteractive.com/harris_poll/index.asp?PID=982.

———. "The Religious and Other Beliefs of Americans 2003." Harris Poll, February 26, 2003. Retrieved April 30, 2009, from http://www.harrisinteractive.com/harris_poll/index.asp?pid=359.

———. "While Most U.S. Adults Believe in God, Only 58 Percent Are 'Absolutely Certain.'" Harris Interactive, October 31, 2006. Retrieved April 24, 2009, from www.harrisinteractive.com/harris_poll/index.asp?PID=707.

Hatch, N. O. *The Democratization of American Christianity*. New Haven, CT: Yale University Press, 1989.

Havely, N. R. *Dante*. Malden, MA: Blackwell, 2007.

Hawkins, P. S. *Dante: A Brief History*. Malden, MA: Blackwell, 2006.

———. *Undiscovered Country: Imagining the World to Come*. New York: Seabury Books, 2009.

Hayes, J. H., and S. Mandell. *The Jewish People in Classical Antiquity*. Louisville, KY: Westminster John Knox Press, 1998.

Heaney, S. *Station Island*. New York: Farrar, Straus and Giroux, 1985.

Hemingway, E., and C. Baker. *Ernest Hemingway, Selected Letters, 1917–1961*. New York: Scribner, 1981.

Henderson, B. "Open Letter to Kansas School Board." *Church of the Flying Spaghetti Monster*. Retrieved May 28, 2009, from www.venganza.org/about/open-letter/.

Henninger, J. "Pre-Islamic Bedouin Religion." In *Studies on Islam*, ed. M. L. Swartz, 3–22. New York: Oxford University Press, 1981.

Himmelfarb, M. *Ascent to Heaven in Jewish and Christian Apocalypses*. Oxford: Oxford University Press, 1993.

Hirschman, J. E., and J. E. Cole. *I Love You All the Time*. Atlanta: Cookie Bear Press, 2001.

Hitchens, C. *God Is Not Great: How Religion Poisons Everything*. New York: Grand Central Publishing, 2007.

Hoffman, L. A. *The Journey Home: Discovering the Deep Spiritual Wisdom of the Jewish Tradition*. Boston: Beacon Press, 2002.

Hogan, R. M. *Dissent from the Creed: Heresies Past and Present*. Huntington, IN: Our Sunday Visitor, 2001.

Holloway, M. *Utopian Communities in America, 1680–1880*. New York: Dover, 1966.

Holy Land Earth. "Suggested Uses for Holy Land Earth." *Holy Land Earth LLC*. Retrieved April 24, 2009, from http://www.holylandearth.com/uses.asp.

The Holy Bible (King James Version). New York: Cambridge University Press, 2000.

The Holy Bible: Containing the Old and New Testaments with the Apocryphal/Deuterocanonical Books: New Revised Standard Version. New York: Oxford University Press, 1989.

Hughes, R. *Heaven and Hell in Western Art*. London: Weidenfeld & Nicolson, 1968.

Ibn 'Alawi, a.-S. M. "The Hadith of Isra' and Mi'raj." In *Islamic Doctrines and Beliefs*. Damascus, Syria: ISCA, 1999.

Ilg, R. E., and A. Clinton. "Strong Job Growth Continues, Unemployment Declines in 1997." *Monthly Labor Review* 121, no. 2 (1998): 48–68.

Inayatullah, S. "Pre-Islamic Thought." In *A History of Muslim Philosophy*, ed. M. M. Sharif, 1:126–35. Kempten, Pakistan: Philosophical Congress, 1963.

IntelCenter. *IntelCenter Terrorism Incident Reference (TIR): Pakistan: 2000–2007*. Alexandria, VA: Tempest, 2008.

Irenaeus. "Against Heresies." *New Advent*. Retrieved May 14, 2009, from http://www.newadvent.org/fathers/0103531.htm.

Irving, W. *Rip van Winkle*. Utrecht, Netherlands: The Catharijne Press, 1987.

Islam Awareness.net. "Miracle of Al-Isra and Al-Miraj." *IslamAwareness.net*. Retrieved May 20, 2009, from http://www.islamawareness.net/Isra/miracle.html.

Israel, S. *Charge!: History's Greatest Military Speeches*. Annapolis, MD: Naval Institute Press, 2007.

Jacobovici, S., and C. Pellegrino. *The Jesus Family Tomb: The Discovery, the Investigation, and the Evidence That Could Change History*. New York: HarperOne, 2007.

Jansen, K. *Ketamine: Dreams and Realities*. Sarasota, FL: Multidisciplinary Association for Psychedelic Studies, 2000.

———. "Neuroscience, Ketamine, and the Near-Death Experience: The Role of Glutamate and the NMDA Receptor." In *The Near-Death Experience: A Reader*, ed. L. W. Bailey and J. Yates, 265–82. New York: Routledge, 1996.

Jensen, R. M. *Understanding Early Christian Art*. London: Routledge, 2000.

Josephus. "The Jewish War." In *Jewish Life and Thought Among Greeks and Romans*, ed. L. H. Feldman and M. Reinhold, 248–51. Minneapolis: Augsburg Fortress, 1996.

Jung, C. G. *Memories, Dreams, Reflections*. New York: Vintage Books, 1989.

K'Meyer, T. E. *Interracialism and Christian Community in the Postwar South: The Story of Koinonia Farm*. Charlottesville: University Press of Virginia, 2000.

Kamenetz, R. *The Jew in the Lotus: A Poet's Rediscovery of Jewish Identity in Buddhist India*. Northvale, NJ: Jason Aronson, 1998.

Keller, H. *My Religion*. New York: Swedenborg Foundation, 1956.

Kelley, D. H., and E. F. Milone. *Exploring Ancient Skies: An Encyclopedic Survey of Archaeoastronomy*. New York: Springer, 2005.

Khalidi, T. *The Qur'an*. London: Penguin Books, 2008.

Kirsch, J. *A History of the End of the World: How the Most Controversial Book in the Bible Changed the Course of Western Civilization*. San Francisco: HarperSanFrancisco, 2006.

Kittelson, J. M. *Luther the Reformer: The Story of the Man and His Career*. Minneapolis: Augsburg, 1986.

Kleinfield, N. R. "Doors Closed, Kennedys Offer Their Farewells." *New York Times*, July 24, 1999.

Koinonia Partners. "Con Browne's Talks in Late February 2006, as Recorded by Ann Karp." Retrieved April 26, 2009, from http://www.koinoniapartners.org/History/oralhistory/Con_Browne.html.

———. "Koinonia Memories from Ms. Georgia Solomon as Told to Ann Karp." Retrieved April 25, 2009, from http://www.koinoniapartners.org/History/oralhistory/Georgia_Solomon.html.

Kraemer, J. L. *Maimonides: The Life and World of One of Civilization's Greatest Minds*. New York: Doubleday, 2008.

Laderman, G. *The Sacred Remains: American Attitudes toward Death, 1799–1883*. New Haven, CT: Yale University Press, 1999.

LaHaye, T. F., and J. B. Jenkins. *Left Behind: A Novel of the Earth's Last Days*. Wheaton, IL: Tyndale House, 1995.

Lamb, C. *Essays of Elia: To Which Are Added Letters, and Rosamund, a Tale*. Paris: Baudry's European Library, 1835.

Lamm, M. *The Jewish Way in Death and Mourning*. Middle Village, NY: Jonathan David Publishers, 2000.

Lang, B. *Sacred Games: A History of Christian Worship*. New Haven, CT: Yale University Press, 1997.

Lansing, R. H., ed. *The Dante Encylopedia*. New York: Garland, 2000.

Latino, J. N. "Year of St. Paul Offers Indulgences." MississippiCatholic.com, November 14, 2008. Retrieved May 13, 2009, from http://www.mississippicatholic.com/categories/diocese/2008/111408/indulgences.html.

Le Goff, J. *The Birth of Purgatory*. Chicago: University of Chicago Press, 1984.

Le Guin, U. K. "Paradises Lost." In *The Birthday of the World and Other Stories*, 213–48. New York: HarperCollins, 2002.

Levenson, J. D. *Resurrection and the Restoration of Israel: The Ultimate Victory of the God of Life*. New Haven, CT: Yale University Press, 2006.

Levine, L. I. *The Ancient Synagogue: The First Thousand Years*. New Haven, CT: Yale University Press, 2005.

Lewis, B., and B. E. Churchill. *Islam: The Religion and the People*. Upper Saddle River, NJ: Wharton School Publishing, 2008.

Lewis, C. S. *The Great Divorce*. New York: HarperCollins, 2001.

———. *The Last Battle*. New York: HarperCollins, 1994.

———. *The Lion, the Witch and the Wardrobe; a Story for Children*. New York: Macmillan, 1950.

———. *Surprised by Joy; The Shape of My Early Life*. London: G. Bles, 1955.

LightPlanet.com. "Temples: A Virtual Tour of Celestial Rooms." LightPlanet.com. Retrieved May 16, 2009, from http://www.lightplanet.com/mormons/temples/celestial_room.html.

Linder, R. D. *The Reformation Era*. Westport, CT: Greenwood Press, 2008.

Lindskoog, K. A. "Botticelli's 'Primavera' and Dante's 'Purgatory.' " In *Dante's Divine Comedy: Purgatory*. Macon, GA: Mercer University Press, 1997.

Lings, M. *Muhammad: His Life Based on the Earliest Sources*. London: Islamic Texts Society, 1983.

Lockhart, R. B. *Halfway to Heaven: The Hidden Life of the Sublime Carthusians*. New York: Vanguard Press, 1985.

Lockman, Z., and J. Beinin. *Intifada: The Palestinian Uprising Against Israeli Occupation*. Boston: South End Press, 1989.

Long, L. A. *Rehabilitating Bodies: Health, History, and the American Civil War*. Philadelphia: University of Pennsylvania Press, 2004.

Longfellow, H. W. "The Reaper and the Flowers." In *The Poetical Works of Henry Wadsworth Longfellow*, 1:22–23. Boston: Houghton Mifflin, 1986.

Lotz, A. G. *Heaven: My Father's House*. Nashville: W Publishing Group, 2001.

Lull, T. F. "Indulgences." In *The Westminster Dictionary of Christian Theology*, ed. A. Richardson and J. Bowden, 295–97. Philadelphia: Westminster Press, 1983.

Luther, M. "Preface to the Revelation of St. John." In *Luther's Works*, vol. 35, ed. E. T. Bachmann. Philadelphia: Fortress, 1960.

———. *Works of Martin Luther, with Introduction and Notes*. Philadelphia: A. J. Holman, 1915.

Luttikhuizen, G. P. E., ed. *Paradise Interpreted: Representations of Biblical Paradise in Judaism and Christianity*. Leiden: E. J. Brill, 1999.

Lynch, T. *The Undertaking: Life Studies from the Dismal Trade*. New York: W. W. Norton, 1997.

Lyons, L. "Paranormal Beliefs Come (Super)Naturally to Some." *Gallup.com*, November 1, 2005. Retrieved May 18, 2009, from http://www.gallup.com/poll/19558/Paranormal-Beliefs-Come-SuperNaturally-Some.aspx.

"Lysippos." *The Grove Encyclopedia of Classical Art and Architecture*. Retrieved April 23, 2009, from http://www.oxfordreference.com/views/ENTRY.html?subview=Main&entry=t231.e0588.

MacDougall, E. B. *Medieval Gardens*. Washington, DC: Dumbarton Oaks Research Library and Collection, 1986.

Madigan, K., and J. D. Levenson. *Resurrection: The Power of God for Christians and Jews*. New Haven, CT: Yale University Press, 2008.

Magness, J. *The Archaeology of Qumran and the Dead Sea Scrolls*. Grand Rapids, MI: William B. Eerdmans, 2002.

Maimonides, M. *Mishneh Torah: The Book of Knowledge*. Jerusalem: Boys Town Jerusalem Publishers, 1965.

Marcantel, M. "Thank God It's Doomsday!" *The Simpsons*. Fox, 2005.

Martin, J. *My Life with the Saints*. Chicago: Loyola Press, 2006.

Martin, W. C. *A Prophet with Honor: The Billy Graham Story*. New York: W. Morrow, 1991.

Marty, M. E. *Pilgrims in Their Own Land: 500 Years of Religion in America*. Boston: Little, Brown, 1984.

Mather, C. *Manuductio ad Ministerium: Directions for a Candidate of the Ministry*. New York: Columbia University Press, 1938.

Mattson, I. "RE: Statements Made by ISNA Canada Secretary General Regarding Dr. Laleh Bakhtiar's Qur'an Translation." The Islamic Society of North America, October 24, 2007. Retrieved May 26, 2009, from http://www.isna.net/articles/Press-Releases/PUBLIC-STATEMENT.aspx.

McDannell, C., and B. Lang. *Heaven: A History*. New York: Vintage Books, 1990.

McGinn, B. *The Presence of God: A History of Western Christian Mysticism*. New York: Crossroad, 1992.

McGrath, A. E. *A Brief History of Heaven*. Malden, MA: Blackwell, 2003.

———. *Reformation Thought: An Introduction*. Oxford: Blackwell, 1999.

McGuire, B. P. "A Lost Clairvaux Exemplum Found: The Liber Visionum Et Miraculorum Compiled Under Prior John of Clairvaux." *Analecta Cisterciensia* 39, no. 1 (January–June 1983): 27–62.

McNeill, J. T., and H. M. Gamer. *Medieval Handbooks of Penance: A Translation of the Principal "Libri Poenitentiales" and Selections from Related Documents*. New York: Columbia University Press, 1990.

McPherson, J. M. *The Atlas of the Civil War*. New York: Macmillan, 1994.

Meacham, J. "Pilgrim's Progress." *Newsweek*, August 14, 2006.

Meacham, J., and L. Miller. "Everything Old Is New Again." *Newsweek*, May 5, 2008.

Mechthild of Magdeburg and F. J. Tobin. *The Flowing Light of the Godhead*. New York: Paulist Press, 1998.

Medici, L. de, and L. Cavalli. *Opere: A Cura di Luigi Cavalli*. Naples: F. Rossi, 1970.

Merton, T. "Introduction." In *Religion in Wood: Masterpieces of Shaker Furniture*, ed. E. D. Andrews and F. Andrews, 7–18. Bloomington: Indiana University Press, 1966.

———. *The Wisdom of the Desert*. New York: New Direction Books, 1960.

Meyer, M. A. *Response to Modernity: A History of the Reform Movement in Judaism*. New York: Oxford University Press, 1988.

Michelangelo Buonarroti. *The Creation of Adam*. Vatican City, Sistine Chapel, 1508–12.

Miles, J. *God: A Biography*. New York: Alfred A. Knopf, 1995.

Miller, J. L. *Dante and the Unorthodox: The Aesthetics of Transgression*. Waterloo, ON: Wilfrid Laurier University Press, 2005.

Miller, L. "The Gospel of Prothero." *Newsweek*, March 12, 2007.

———. "Is Obama the Antichrist?" *Newsweek*, November 15, 2008.

———. "Life in Solitary." *Newsweek*, June 20, 2005.

———. "Religion: The Age of Divine Disunity—Faith Now Springs from a Hodgepodge of Beliefs." *Wall Street Journal*, February 10, 1999.

———. "Why We Need Heaven." *Newsweek*, August 12, 2002.

Miller, L., and A. Murr. "Jesus and Witches." *Newsweek*, October 28, 2008. Retrieved April 20, 2009, from http://www.newsweek.com/id/166215.

Miller, L., and R. Wolffe. "Finding His Faith." *Newsweek*, July 12, 2008.

———. "I Am a Big Believer in Not Just Words, But Deeds and Works." *Newsweek*, July 21, 2008.

Miller, R. M., H. S. Stout, and C. R. Wilson, eds. *Religion and the American Civil War*. New York: Oxford University Press, 1998.

Milosz, C. *To Begin Where I Am*. New York: Farrar, Straus and Giroux, 2001.

Mission to Empty Purgatory. "Calculations." MTEP.com. Retrieved May 14, 2009, from http://www.mtep.com/calculations.htm.

Mitchell, J. *A Discourse of the Glory to Which God Hath Called Believers by Jesus Christ Delivered in Some Sermons out of the I Pet. 5 Chap. 10 ver.: Together with an Annexed Letter*. London: Printed for Nathaniel Ponder at the Peacock in the Poultry, 1677.

Mitchell, S. *Gilgamesh: A New English Version*. New York: Free Press, 2004.

Mitford, J. *The American Way of Death*. New York: Simon & Schuster, 1963.

Mohr, R. D. *God and Forms in Plato: The Platonic Cosmology*. Las Vegas: Parmenides Publishing, 2005.

Moody, R. *Life After Life*. New York: Bantam, 1978.

Moody, R. A., and P. Perry. *Life Before Life: Regression into Past Lives*. London: Macmillan, 1990.

Moore, D. W. "Three in Four Americans Believe in Paranormal." In *The Gallup Poll*, 221–22. Lanham, MD: Rowman & Littlefield, 2005.

Moore, R. L. *In Search of White Crows: Spiritualism, Parapsychology, and American Culture*. New York: Oxford University Press, 1977.

Morrison, G., ed. *History of Persian Literature from the Beginning of the Islamic Period to the Present Day*. Leiden: E. J. Brill, 1981.

Moulting Mantis Library. "Apocalypse of John: A Poem of Terrible Beauty," trans. Leonard L. Thompson, www.moultingmantis.org, 2003.

Muessig, C., and A. Putter. *Envisaging Heaven in the Middle Ages*. London: Routledge, 2007.

Newberg, A., and M. R. Waldman. *How God Changes Your Brain: Breakthrough Findings from a Leading Neuroscientist*. New York: Ballantine Books, 2009.

Newton, J. "Catholic Church (North America)." In *The Encyclopedia of Cremation*, ed. D. J. Davies and L. H. Mates, 112. Burlington, VT: Ashgate, 2005.

Niebuhr, R. *Christianity and Power Politics*. New York: Charles Scribner's Sons, 1940.

NIV Study Bible, 10th Anniversary Edition. Grand Rapids, MI: Zondervan Publishing House, 1995.

Noll, M. A. *A History of Christianity in the United States and Canada*. Grand Rapids, MI: William B. Eerdmans, 1992.

Nuland, S. B. *How We Die: Reflections on Life's Final Chapter*. New York: Alfred A. Knopf, 1994.

———. *Maimonides*. New York: Nextbook/Schocken, 2005.

Nydell, M. K. *Understanding Arabs: A Guide for Modern Times*. Yarmouth, ME: Intercultural Press, 2006.

Oberman, H. A. *Luther: Man Between God and the Devil*. New Haven, CT: Yale University Press, 2006.

Olson, D. C. "1 Enoch." In *Eerdmans Commentary on the Bible*, ed. J. D. G. Dunn and J. W. Rogerson, 904–41. Grand Rapids, MI: William B. Eerdmans, 2003.

Origen. *Origen on Prayer*. Grand Rapids, MI: Christian Classics Ethereal Library, 2001.

Paludan, P. S. "Religion and the American Civil War." In *Religion and the American Civil War*, ed. R. M. Miller, H. S. Stout, and C. R. Wilson. New York: Oxford University Press, 1998.

Pareles, J. "Pop and Jazz Guide." *New York Times*, June 29, 2001.

———. "A Rock Star's Struggle Where Militant Islam Rules." *New York Times*, July 17, 2003.

Patterson, B. B. "Appalachian Religious Music." In *Encyclopedia of Religion in the*

South, ed. S. S. Hill and C. H. Lippy, 69–73. Macon, GA: Mercer University Press, 2005.

Penton, M. J. *Apocalypse Delayed: The Story of Jehovah's Witnesses*, 2nd ed. Toronto: University of Toronto Press, 1997.

Pestana, C. G. *Quakers and Baptists in Colonial Massachusetts*. Cambridge: Cambridge University Press, 1991.

Peters, F. E. *Mecca: A Literary History of the Muslim Holy Land*. Princeton, NJ: Princeton University Press, 1994.

Petersen, J. R. "Virgins in Paradise: The Strange Erotic Visions of a Suicide Bomber." *Playboy*, April 2002.

———. "Many Americans Say Other Faiths Can Lead to Eternal Life." *PewForum. org*. Retrieved April 13, 2009, from http://pewforum.org/docs/?DocID=380.

———. "Pentecostalism." *PewForum.org*, Retrieved May 18, 2009, from http://pewfo rum.org/docs/?DocID=140.

Pew Forum on Religion and Public Life. *U.S. Religious Landscape Survey: Religious Affiliation*. Washington, DC, 2008.

———. *U.S. Religious Landscape Survey: Religious Beliefs and Practices*. Washington, DC, 2008.

Pew Research Center. *Muslim Americans: Middle Class and Mostly Mainstream*. Washington, DC, 2007.

Pew Research Center for People and the Press. "Many Americans Uneasy with Mix of Religion and Politics." *People-Press.org*, Retrieved April 22, 2009, from http:// people-press.org/report/?pageid=1084.

———. *2004 Political Landscape*. Washington, DC, 2003.

Phan, P. C. "Roman Catholic Theology." In *The Oxford Handbook of Eschatology*, ed. J. L. Walls, 215–32. Oxford: Oxford University Press, 2008.

Phelps, E. S. *The Gates Ajar*. Boston: Fields, Osgood, 1869. Revised edition.

Pilla, A. "Building the City of God." *EcoCity Cleveland*, Retrieved April 14, 2009, from http://www.ecocitycleveland.org/smartgrowth/cornfields/city_of_god.html.

Pindar, A. *The Complete Odes*. Translated by A. Verity. With an introduction and notes by S. Instone. Oxford: Oxford University Press, 2007.

Piper, D., and C. B. Murphey. *Heaven Is Real: Lessons on Earthly Joy from the Man Who Spent 90 Minutes in Heaven*. New York: Penguin Books, 2007.

———. *90 Minutes in Heaven: A True Story of Death and Life*. Grand Rapids, MI: Revell, 2004.

Placher, W. C. *A History of Christian Theology: An Introduction*. Philadelphia: Westminster Press, 1983.

Plato. *Phaedo*. Stillwell, KS: Digireads.com, 2006.

Pope, H. "The Kingdom of God." In *The Catholic Encyclopedia*, ed. C. G. Herbermann et al., 8:646–47. New York: The Encyclopedia Press, 1913.

Porten, B. "Exile, Babylonian." In *Encyclopaedia Judaica*, ed. M. Berenbaum and F. Skolnik, 6:608–11. Detroit: Macmillan Reference USA, 2007.

Price, S. R. F. *Rituals and Power: The Roman Imperial Cult in Asia Minor.* Cambridge: Cambridge University Press, 1984.

Prothero, S. R. *American Jesus: How the Son of God Became a National Icon.* New York: Farrar, Straus and Giroux, 2003.

———. *Religious Literacy: What Every American Needs to Know—and Doesn't.* San Francisco: HarperSanFrancisco, 2007.

Purcell, M. *Papal Crusading Policy: The Chief Instruments of Papal Crusading Policy and Crusade to the Holy Land from the Final Loss of Jerusalem to the Fall of Acre 1244–1291.* Leiden: E. J. Brill, 1975.

Queen, E. L., S. R. Prothero, and G. H. Shattuck. "Spiritualism." In *Encyclopedia of American Religious History*, 2:704. New York: Facts on File, 2001.

Raboteau, A. J. *Slave Religion: The "Invisible Institution" in the Antebellum South.* New York: Oxford University Press, 1978.

Ralls, M. "What Can We Say About the Afterlife?" *The Christian Century*, December 14, 2004.

Raphael. *Madonna of the Meadows.* Vienna, Kunsthistorisches Museum: Oil tempera on wood, 1505/1506.

Raphael, S. P. *Jewish Views of the Afterlife.* Northvale, NJ: Jason Aronson, 1994.

Rappaport, U. "Jason." In *Encyclopaedia Judaica*, ed. M. Berenbaum and F. Skolnik, 11:90. Detroit: Macmillan Reference USA, 2007.

Ratzinger, J. "'In the Presence of the Angels I Will Sing Your Praise': The Regensburg Tradition and the Reform of the Liturgy." *Adoremus Bulletin*, October–December 1996. Retrieved May 18, 2009, from http://www.adoremus.org/10–12–96-Ratzi .html.

Rauschenbusch, W. *Christianity and the Social Crisis.* New York: Macmillan, 1907.

Raza, A. "The Petition for Professor Khaled Abou El Fadl to Come and Speak at NYU." Facebook. Retrieved April 27, 2009, from http://www.facebook.com/ group.php?sid=48cb31e9b470b4f86cb7becba3dbe8c0&gid=33119669676&ref= search.

Renick, T. *Aquinas for Armchair Theologians.* Louisville: Westminster John Knox Press, 2002.

Roberts, A., and J. Donaldson, eds. *The Writings of the Apostolic Fathers.* London: Elibron Classics, 2005.

Robinson, C. F. *Islamic Historiography.* Cambridge: Cambridge University Press, 2003.

Robinson, M. *Gilead.* New York: Farrar, Straus and Giroux, 2004.

Rodinson, Maxime. *Muhammad.* London: Tauris Parke, 2002.

Roman Catholic Diocese of Brooklyn. "Announces Indulgences During Pauline Year."

DioceseofBrooklyn.org, August 5, 2008. Retrieved May 13, 2009, from http://di oceseofbrooklyn.org/default_article.aspx?id=2084.

Roof, W. C. A Generation of Seekers: The Spiritual Journeys of the Baby Boom Generation. San Francisco: HarperSanFrancisco, 1993.

Rosner, F. The Medical Legacy of Moses Maimonides. Hoboken, NJ: Ktav, 1998.

Roth, P. Indignation. Boston: Houghton Mifflin, 2008.

Rothkoff, A. "Mitzvah." In Encyclopaedia Judaica, ed. M. Berenbaum and F. Skolnik, 14:372. Detroit: Macmillan Reference USA, 2007.

Rowling, J. K. Harry Potter and the Philosopher's Stone. London: Bloomsbury, 1997.

Rubens, P. P. The Assumption of the Virgin Mary. Antwerp, Antwerp Cathedral: Oil on board, 1626.

Rubin, U. "Hanîf." In Encyclopaedia of the Qur'ân, ed. J. D. McAuliffe, 2:402. Leiden: E. J. Brill, 2009.

Ruggles, D. F. Islamic Gardens and Landscapes. Philadelphia: University of Pennsylvania Press, 2008.

Rumi, J. a.-D., and C. Barks. The Essential Rumi. San Francisco: Harper, 1995.

Russell, B. A History of Western Philosophy. New York: Simon and Schuster, 1972.

Russell, J. B. A History of Heaven: The Singing Silence. Princeton, NJ: Princeton University Press, 1997.

———. Paradise Mislaid: How We Lost Heaven—and How We Can Regain It. Oxford: Oxford University Press, 2006.

Sanders, E. P. The Historical Figure of Jesus. London: Allen Lane, 1993.

———. Jesus and Judaism. Philadelphia: Fortress Press, 1985.

Scafi, A. Mapping Paradise: A History of Heaven on Earth. Chicago: University of Chicago Press, 2006.

Schaff, P. History of the Christian Church. New York: Charles Scribner's Sons, 1891.

Schalit, A., and L. Matassa. "Elephantine." In Encyclopaedia Judaica, ed. M. Berenbaum and F. Skolnik, 6:311–14. Detroit: Macmillan Reference USA, 2007.

Schalit, A., et al. (2007). "Alexander the Great." In Encyclopaedia Judaica, ed. M. Berenbaum and F. Skolnik, 1:625–27. Detroit: Macmillan Reference USA.

Schillebeeckx, E. For the Sake of the Gospel. New York: Crossroad, 1990.

Schimmel, A. "Some Aspects of Mystical Prayer in Islam." Die Welt des Islams, New Series 2, no. 2 (1952): 112–25.

Schweitzer, A., and W. Lowrie. The Mystery of the Kingdom of God: The Secret of Jesus' Messiahship and Passion. London: A. & C. Black, 1925.

Sebold, A. The Lovely Bones: A Novel. Boston: Little, Brown, 2002.

Segal, A. F. Life After Death: A History of the Afterlife in the Religions of the West. New York: Doubleday, 2004.

Shakespeare, W. Hamlet. London: Cornmarket, 1969.

Sharpe, M., ed. Suicide Bombers: The Psychological, Religious and Other Imperatives.

NATO Science for Peace and Security Series—Human and Societal Dynamics 41 (July 2008). Amsterdam: IOS Press, 2008.

Shriver, M. *What's Heaven?* New York: Golden Books, 1999.

Sife, W. *The Loss of a Pet.* New York: Howell Book House, 1998.

Sigstedt, C. O. *The Swedenborg Epic: The Life and Works of Emanuel Swedenborg.* New York: Record Press, 1971.

Simoons, F. J. *Plants of Life, Plants of Death.* Madison: University of Wisconsin Press, 1998.

Smith, J. I., and Y. Y. Haddad. *The Islamic Understanding of Death and Resurrection.* Oxford: Oxford University Press, 2002.

Smith, M. *The Memoirs of God: History, Memory and the Experience of the Divine in Ancient Israel.* Minneapolis: Fortress Press, 2004.

Southern Baptist Theological Seminary. "What We Believe." Retrieved April 30, 2009, from http://www.sebts.edu/about/what-we-believe/default.aspx.

Sozomenus. "The Ecclesiastical History of Sozomen, Comprising a History of the Church from AD 323 to AD 425." In *A Select Library of Nicene and Post-Nicene Fathers of the Christian Church*, 2nd series, ed. P. Schaff and H. Wace, 2:179–455. New York: Christian Literature Company, 1890.

"Spirit Mediums Outdone. Lively Rappings in the Academy of Music. Dr. Richmond and One of the Fox Sisters Give Exhibitions of Their Skill Before a Remarkably Responsive Crowd—Spiritualism Formally Renounced." *New York Tribune*, October 22, 1888.

Stark, R. *Discovering God: The Origins of the Great Religions and the Evolution of Belief.* New York: HarperOne, 2007.

Stein, J. "A Little Bit of Heaven on Earth." *Los Angeles Times*, December 21, 2007.

Steinberg, A., and F. Rosner. *Encyclopedia of Jewish Medical Ethics: A Compilation of Jewish Medical Law on All Topics of Medical Interest.* Jerusalem: Feldheim, 2003.

Stille, A. "Scholars Are Quietly Offering New Theories of the Koran." *New York Times*, March 2, 2002.

Stone, D. "One Nation Under God?" *Newsweek*, April 7, 2009, http://www.newsweek.com/id/192915.

Stout, H. S. *The New England Soul: Preaching and Religious Culture in Colonial New England.* New York: Oxford University Press, 1986.

Sutton, R. P. *Communal Utopias and the American Experience: Religious Communities 1732–2000.* Westport, CT: Praeger, 2003.

Swan, L. *The Forgotten Desert Mothers.* Mahwah, NJ: Paulist Press, 2001.

Swedenborg, E. *The Delights of Wisdom Respecting Conjugal Love: After Which Follow the Pleasures of Insanity Respecting Scortatory Love.* London: Printed for the Society, 1790.

———. *Heaven and Its Wonders and Hell, Drawn from Things Heard and Seen.* West

Chester, PA: Swedenborg Foundation, 2000.

Swedenborg Foundation. "Illuminating the World of Spirit." *Swedenborg.com*. Retrieved May 14, 2009, from http://www.swedenborg.com/catalog/2005catalog.pdf.

Symonds, J. A. *The Sonnets of Michael Angelo Buonarroti and Tommaso Campanella*. Teddington, UK: The Echo Library, 2007.

Taylor, B. "Leaving Myself Behind." In *Heaven*, ed. R. Ferlo. New York: Seabury Books, 2007.

Taylor, J. E. *The Immerser: John the Baptist Within Second Temple Judaism*. Grand Rapids, MI: William B. Eerdmans, 1997.

Taylor, M. R. "Dealing with Death: Western Philosophical Perspectives." In *Handbook of Death and Dying*, ed. C. D. Bryant, 1:24–33. Thousand Oaks, CA: Sage, 2003.

Tcherikover, V. *Hellenistic Civilization and the Jews*. Philadelphia: Jewish Publication Society of America, 1959.

Teeman, T. "Attention: Towering Intellect at Work." *The Times*, January 24, 2007.

Teibel, A. "Embracing Secular Culture Can Be Risky in Israel." Associated Press, October 5, 2008.

Thompson, L. L. *The Book of Revelation: Apocalypse and Empire*. New York: Oxford University Press, 1990.

Thurston, H. "Simeon Stylites the Elder." In *The Catholic Encyclopedia*, ed. C. G. Herbermann, 13:795. New York: Encyclopedia Press, 1913.

Tillich, P. *Systematic Theology*. Chicago: University of Chicago Press, 1967.

Tocqueville, A. de *The Tocqueville Reader: A Life in Letters and Politics*, ed. O. Zunz and A. S. Kahan. Malden, MA: Blackwell, 2002.

Toner, P. "Limbo." In *The Catholic Encyclopedia*. New York: Robert Appleton Company, 1910. Retrieved May 28, 2009, from http://www.newadvent.org/cathen/09256a.htm.

Trumbower, J. A. *Rescue for the Dead: The Posthumous Salvation of Non-Christians in Early Christianity*. Oxford: Oxford University Press, 2001.

Twain, M. *Extract from Captain Stormfield's Visit to Heaven*. New York: Oxford University Press, 1996.

Tyler, A. F. *Freedom's Ferment; Phases of American Social History to 1860*. Minneapolis: University of Minnesota Press, 1994.

U.S. Census Bureau. "Interim Projections of the Population by Selected Age Groups for the United States and States: April 1, 2000, to July 1, 2030." Retrieved April 14, 2009, from http://www.census.gov/population/projections/SummaryTabB1.pdf.

Underhill, A. L. *The Missing Link in Modern Spiritualism*. New York: Thomas R. Knox, 1885.

"Unhappy America." Leaders. *The Economist*, July 24, 2008.

Unitarian Universalist Association of Congregations. "Universalism." April 18, 2008. Retrieved May 28, 2009, from http://www.uua.org/visitors/ourhistory/6904.shtml.

Valley of Life.com. "In Loving Memory: Johnnie Lane Parker, Sr." ValleyOfLife.com. Retrieved May 18, 2009, from http://www.valleyoflife.com/JohnnieLParkerSr/.

Van Biema, D. "Does Heaven Exist?" *Time*, March 24, 1997.

Van Gogh, V. "Letter to Theo van Gogh, 28 September 1888." *The Letters of Vincent van Gogh*. Retrieved April 25, 2009, from http://www.webexhibits.org/vangogh/letter/18/543.htm.

Vanderhoof, E. W. *Historical Sketches of Western New York*. New York: AMS Press, 1972.

Voll, J. O. *Islam, Continuity and Change in the Modern World*. Syracuse, NY: Syracuse University Press, 1994.

Voll, U., and S. A. Kenel. "Deadly Sins." In *New Catholic Encyclopedia*, 2nd ed., 4:565–67. Detroit: Thomson/Gale, 2003.

Wagner, M. "Shas Pulls Controversial Election Ad." *Jerusalem Post*, March 12, 2006.

Walls, J. L. *Heaven: The Logic of Eternal Joy*. Oxford: Oxford University Press, 2002.

Ward, B. *The Desert Fathers: Sayings of the Early Christian Monks*. London: Penguin Books, 2003.

Waterhouse, G. "Another Early German Account of St. Patrick's Purgatory." *Hermathena* 23 (1933): 114–16.

Watt, W. M. *Muhammad: Prophet and Statesman*. London: Oxford University Press, 1974.

———. *Muhammad at Mecca*. Oxford: Clarendon Press, 1953.

———. *Muhammad at Medina*. Oxford: Clarendon Press, 1956.

Watt, W. M., A. J. Wensinck, et al. "Makka." In *Encylopaedia of Islam*, ed. P. Bearman, T. Bianquis, C. E. Bosworth, E. v. Donzel, and W. P. Heinrichs. Leiden: E. J. Brill, 2009.

Wawrykow, J. P. *The Westminster Handbook to Thomas Aquinas*. Louisville, KY: Westminster John Knox Press, 2005.

Webb, D. *Pilgrims and Pilgrimage in the Medieval West*. London: I. B. Tauris, 1999.

Weinfeld, M. *Normative and Sectarian Judaism in the Second Temple Period*. London: T. & T. Clark, 2005.

Weiss, B. L. *Many Lives, Many Masters*. New York: Simon & Schuster, 1988.

Welch, E. S. *Art in Renaissance Italy, 1350–1500*, new edition. Oxford: Oxford University Press, 2000; first printed in 1997.

Wensinck, A. J., and B. Lewis. "Hadjdj." In *Encyclopaedia of Islam*, ed. P. Bearman, T. Bianquis, C. E. Bosworth, E. v. Donzel, and W. P. Heinrichs, 3:31. Leiden: E. J. Brill, 2009.

Wessels, A. *A Modern Arabic Biography of Muhammad*. Leiden: E. J. Brill, 1972.

White, L. M. "Apocalyptic Literature in Judaism and Early Christianity." PBS.org. Retrieved May 1, 2009, from www.pbs.org/wgbh/pages/frontline/shows/apocalypse/primary/white.html.

———. *From Jesus to Christianity.* San Francisco: HarperSanFrancisco, 2004.

White, S. J. *Foundations of Christian Worship.* Louisville, KY: Westminster John Knox Press, 2006.

White, W. M. *Emanuel Swedenborg: His Life and Writings.* London: Simpkin, Marshall, 1867.

Whitney, H., and J. Barnes. "The Mormons." *Frontline.* PBS, broadcast April 30 and May 1, 2007.

Williams, R. H. "An Illustration of Historical Inquiry: Histories of Jesus and Matthew 1:1–25." In *Handbook of Early Christianity,* ed. A. J. Blasi, P.-A. Turcotte, and J. Duhaime, 105–24. Walnut Creek, CA: AltaMira Press, 2002.

Wills, G. *Saint Augustine.* New York: Lipper/Viking, 1999.

Winseman, A. L. "Eternal Destinations: Americans Believe in Heaven, Hell." Gallup, Inc., May 25, 2004. Retrieved April 13, 2009, from http://www.gallup.com/poll/11770/Eternal-Destinations-Americans-Believe-Heaven-Hell.aspx.

Woodward, B. "In Hijacker's Bags, a Call to Planning, Prayer and Death." *Washington Post,* September 28, 2001.

Work of the Chariot. "The Measure of the Divine Body." Work of the Chariot. Retrieved May 18, 2009, from http://www.workofthechariot.com/TextFiles/Translations-ShirQoma.html.

Wright, J. A. *What Makes You So Strong?: Sermons of Joy and Strength from Jeremiah A. Wright, Jr.* Edited by J. K. Ross. Valley Forge, PA: Judson Press, 1993.

Wright, N. T. *Jesus and the Victory of God.* Minneapolis: Fortress Press, 1996.

———. *The Resurrection of the Son of God.* Minneapolis: Fortress Press, 2003.

———. *Surprised by Hope: Rethinking Heaven, the Resurrection, and the Mission of the Church.* New York: HarperOne, 2008.

Yinger, K. L. *Paul, Judaism, and Judgment According to Deeds.* Cambridge: Cambridge University Press, 1999.

Young, B. *Discourses of Brigham Young: Second President of the Church of Jesus Christ of Latter-Day Saints.* South Salt Lake City, Utah: Deseret Book Co., 1971.

Zaleski, C. *Otherworld Journeys: Accounts of Near-Death Experience in Medieval and Modern Times.* New York: Oxford University Press, 1987.

Zaleski, C., and P. Zaleski. *The Book of Heaven: An Anthology of Writings from Ancient to Modern Times.* New York: Oxford University Press, 2000.

The Zohar. Trans. H. Sperling and M. Simon. London: Soncino Press, 1934.

INDEX